AF601903

Bioluminescent Marine Invertebrates

THE AUTHOR

Dr. Ramasamy Santhanam, PhD, is the former Dean of the Fisheries College and Research Institute (FC and RI), Tamil Nadu Dr. J. Jayalalithaa Fisheries University, Thoothukudi, India. He has about 50 years of teaching and research experience in marine/fisheries sciences. He has served as a fisheries expert for various government and nongovernment organizations in India. Dr Santhanam has the credit of being authored books with internationally reputed publishers like Springer Verlag, CRC, Academic Press, Apple Academic Press, Nova Science and Bentham Books and he has so far published 35 books on various aspects of marine life, fisheries environment, fisheries biology, and aquaculture. His two titles have been identified by the World Book Authority as "Best Marine Biology Books of All time ". He was a member of the American Fisheries Society, United States; World Aquaculture Society, United States; Global Fisheries Ecosystem Management Network (GFEMN), United States; and the IUCN's Commission on Ecosystem Management, Switzerland.

Bioluminescent Marine Invertebrates

– Author –

Ramasamy Santhanam

2025

Daya Publishing House®

A Division of

Astral International Pvt. Ltd.

New Delhi – 110 002

ISBN: 9789359194141

Published by : **Daya Publishing House®**
A Division of
Astral International Pvt. Ltd.
– ISO 9001:2015 Certified Company –
4736/23, Ansari Road, Darya Ganj
New Delhi-110 002
Ph. 011-43549197, 23278134
E-mail: info@astralint.com
Website: www.astralint.com

TAMIL NADU Dr.J.JAYALALITHAA FISHERIES UNIVERSITY
Dr. M.G.R. FISHERIES COLLEGE AND RESEARCH INSTITUTE
THALAINAYERU - 614 712, NAGAPATTINAM, INDIA

Mobile : 9345718579
E-mail : deanfcritnayeru@tnfu.ac.in

Dr.S.Balasundari, M.F.Sc., Ph.D.
Dean i/c.

29.08.2023

I am very much pleased to write this Foreword for the title "Bioluminescent Marine Invertebrates" by Dr. Ramasamy Santhanam, my former teacher and Ex-Dean of the Fisheries College & Research Institute, Thoothukudi, India. He is well known for his enthusiasm in sharing his knowledge for the benefit of students. This title, a companion volume of his other published titles viz. "Bioluminescent Marine Fishes" and "Bioluminescent Marine Plankton" deal with the emission of "cold living light" of various constituents of marine invertebrates; and its potential uses in the therapeutical and biotechnological applications.

It is my hope and expectation that this book when published will provide an effective learning experience as a textbook and as a referenced source for all Marine Biology/Fisheries Science students and teachers.

I congratulate the author for his timely contribution.

S. Balasundari
29/8/2023
DEAN i/c
Dr. MGR Fisheries College and Research Institute,
Thalainayeru, Nagapattinam - 614 712.

Preface

Bioluminescence, the "cold living light" or the "cold fire of the sea" is extremely common in all oceans at all depths. However, this phenomenon is nearly absent in freshwater, with the exception of a freshwater limpet, *Latia neritoides*. Further, there are more bioluminescent life forms in the sea, than on land. The marine invertebrates such as the luminescent marine sponges, sea pens, worms, shrimps, squids, and echinoderms exert greater impact on the commercial fisheries through their food chain. Recent research has shown that the natural products of the bioluminescent marine invertebrates could be great use in biotechnological and therapeutic applications. For instance, application of " Bioluminescence imaging" has grown tremendously in the past decade and it has significantly contributed to the core conceptual advances in biomedical research. This technology has provided valuable means for monitoring of different biological processes for immunology, oncology, virology and neuroscience. Bioluminescence imaging has also been successfully used to monitor infections caused by various microorganisms, in particular bacteria.

Though a few books are presently available on Bioluminescent Marine Life *viz.* "Bioluminescent Marine Plankton" and "Biology and Ecology of Bioluminescent Marine Fishes" (by this author), a comprehensive volume dealing with the "Bioluminescent Marine Invertebrates " has not so far been published. This publication, the first of its kind, would answer this long felt need. It deals mainly with the Biology and Ecology of Bioluminescent Nektonic and Benthic Marine Invertebrates. It is hoped that the present publication when brought out would be of great use as a standard text-cum-reference for teachers, students and researchers of various disciplines such as Marine Biology, Fisheries Science, Life Sciences and Environmental Sciences; and as a valuable reference for libraries of colleges and universities.

Ramasamy Santhanam

Contents

Chapter 1

Introduction

In around 350 BC, Greek philosopher Aristotle was the first to observe bioluminescence as a type of "cold" light – in that it does not produce heat. He discovered on the self-luminosity in about 180 marine species. Nearly 2,500 years ago,the Greek philosopher Aniximenes found that light emanated from water when struck by an oar (Stokes, https://www.csmonitor.com/Science/2012/0906/On-ocean-floor-a-shrimp-that-vomits-light).On January 6, 1832 when Charles Darwin was aboard the Beagle just off the coast of Tenerife, he saw bioluminescent sea creatures, flickering light in response to physical disruption. Amazingly bioluminescence has even played a part in warfare. As per available report, the bioluminescent organisms helped in the sinking of the last German U-boat during World War One, in November 1918. When the submarine sailed through a bioluminescent bloom, it left a glowing wake which was tracked by the allies. Further, this bioluminescence exhibited a protective role also. Immediately after one of the battles of the American Civil War at Shiloh (southwestern Tennessee, USA), the wounds of some of the injured soldiers were glowing and such wounds healed more quickly and cleanly. It was subsequently known that this glow was due to the light produced by a soil-dwelling bacterium, *Photorhabdus luminescens* which released antimicrobial compounds and thus protected the soldiers from further infection (https://theconversation.com/what-is-bioluminescence-and-how-is-it-used-by-humans-and-in-nature-100472). Some interesting incidents associated with marine bioluminescence observed by earlier workers include: formation of bioluminescent milky seas due to huge populations of luminous bacteria; formation of red tide due to luminescent dinoflagellates strikes in Monterey Bay; observation of a hydrozoan siphonophore using its red light to lure fish to its tentacles; squid changing its color of its luminescence to match sunlight and moonlight; and deep sea crustacean shrimps which were sending out coded messages to their own species during the time of mating; (Haddock, http://photobiology.info/Haddock.html; https://biolum.eemb.ucsb.edu/organism/)

Optical Phenomena

At present, there are many terms describing optical phenomena which are all confused with bioluminescence. Among these terms, the chemiluminescence, fluorescence and phosphorescence are common and they are nothing but an excitation of a material in such a way that light is emitted for each. The characteristics of these three phenomena and their relationship with bioluminescence are given below:

1. **Chemiluminescence:** It is an excitation and the light so produced is the result of a chemical reaction, as in glow-sticks. Bioluminescence can also be considered as a subset of chemiluminescence. This is due to the light produced when such a chemical reaction occurs within a bioluminescent organism.
2. **Fluorescence:** In this case, the excitation occurs when a material is hit by an incident light of particular wavelength. In this case, the molecules are not stable in this excited state and as they came back to their ground state, the energy is almost immediately re-emitted in the form of a photon of a longer wavelength. For this phenomenon, the popular green-fluorescent protein (GFP) from the hydromedusa Aequorea may be considered as an example. Before it is giving a green light, it is known to be excited by blue light. Keeping this in view, like chemiluminescence, bioluminescence is also related to fluorescence, in that many luminous molecules may also be caused to fluoresce. Other examples of fluorescent molecules found in nature are chlorophyll and phycobiliproteins.
3. **Phosphorescence:** In those days, this phenomenon was used as a synonym for sea-water bioluminescence, which is however nothing to do with the true bioluminescence as the scientific meaning of phosphorescence is quite different. Phosphorescence is also similar to fluorescence in that excitation occurs when the material is hit by light. However in this phenomenon, the molecules, are somewhat stable in their excited state, and re-release the energy gradually over time, resulting in a dim but steady glow. The most familiar example for phosphorescence is the glow-in-the-dark stickers. This phenomenon is also seen naturally in some minerals and photosynthetic systems (Haddock, http://photobiology.info/Haddock.html)

Bioluminescence

Bioluminescence is a form of chemiluminescence, producing light as a result of the oxidation of a compound called luciferin by an enzyme called luciferase.

This light is usually blue in colour because this is the light that travels best through the water column. However this light may vary from nearly violet to green-yellow (very occasionally red), and is emitted in three different ways as detailed below:

1. In some species, the light actually appears to be vomited from the animal and is invariably in the form of secreted mucus and this type of luminescence is called extracellular luminescence. In cases of extracellular luminescence, the luminous secretion is discharged into the ambient water. The light-producing glands in this type are either unicellular or multicellular structures which are distributed throughout the body or restricted to definite areas (http://ecoursesonline.iasri.res.in/mod/page/view.php?id=86809)
2. In some other species, the light is emitted by specialized cells called photocytes which may be grouped into complicated lensed structures called photophores looking very much like eyes, except that light comes out instead of going in; This type of intracellular luminescence is also widespread among the marine animals. In *Nocitluca,* the light production is performed by granules present in the periphery of the cell. In hydromedusae, the luminous cells are found grouped beneath the endoderm of marginal canals and in sea pens the luminous cells are in the endoderm of tentacles. Intracellular luminescence is of also wide occurrence among crustaceans such as shrimps and euphausiids. The photophores of certain pelagic caridean and penaeid shrimps are distributed over the appendages (http://ecoursesonline.iasri.res.in/mod/page/view.php?id=86809)
3. In a few species, the light emission is due to colonies of bioluminescent bacteria which are living in black pouches. Though these bacteria can glow continuously, the pouches can open and close, allowing the animal to control the light emission.

Mysteries of Bioluminescence

1. The presence of large number of small light organs on the lower surface of many fish, shrimp, and squid is particularly to hide the silhouette of the animal when it's viewed from below. This is called counterillumination.
2. The stalked light organs of deep-sea fish are believed to be lures for prey and in some cases they help in searching for potential mates. Further, light organs present under and over the eyes of certain fish serve as searchlights
3. Some deep-sea animals use their bioluminescence to communicate with each other.

Bioluminescent Marine Organisms

The seas and oceans are the home for the vast majority of luminous organisms which include dinoflagellates, radiolarians, hydroids, jellyfishes, alcyonarians, ctenophores, bryozoans, polychaetes, brittle stars, many crustaceans, gastropds, bivalves, cephalopods, prochordates, gastropods,bivalves, cephalopods, prochordates, fishes, *etc.*, Bioluminescence is nearly absent in freshwater taxa,with the exception of one gastropod species *viz.*, the limpet, *Latia neritoides* which is endemic to the North Island of New Zealand. Moraes, *et al.* (2021) and Oba *et al.* (2017) reported that bioluminescence is present in about 800 genera and probably over10000 species within 16 marine phyla.

Distribution of Bioluminescence

Light-producing animals are present in a wide variety of environments from polar to tropical regions and they are of wide occurrence from the surface of the sea to abyssal depths. In the tropical waters, bioluminescence is more common than in higher latitudes(http://ecoursesonline.iasri.res.in/mod/page/view.php?id=86809)While the bioluminescence has been reported to be very common in the water column of the ocean, it is believed to be less common on coral reefs and other places near the shore. (Johnsen, https://oceanexplorer.noaa.gov/explorations/15biolum/background/biolum/biolum.html). In the dimly-lit midwater zone at 100-1000 m. depth where the biologically produced bioluminescent light can outshine the existing light which has filtered down from the surface. It has been estimated that in this zone, about 90 per cent of the fish and crustacean species are capable of making their bioluminescent light, and the numbers for gelatinous zooplankton such as jellyfish, siphonophores, comb jellies are said to be even higher in this zone. On the other hand, about 10-20 per cent of seafloor dwellers are only bioluminescent and much less was known about the bioluminescence in organisms living close to the sea floor. Most of the benthic organisms are most sensitive to blue-green light between the emission wavelengths 470 - 497nm. Although fewer benthic species produce light than in the middle depths of the ocean, the sea floor is believed to be much brighter than upper depths. Bioluminescence in the sea floor is like glowing rain with big flashes, due to the presence of several species of coral, sea anemones, and deep-sea shrimp which often vomit bioluminescent chemicals into the water surrounding it. Further, in the sea floor, most of the organisms glow blue, except for a family of corals known as pennatulaceans, which produce green light (https://www.science.org/content/article/vomiting-shrimp-and-other-deep-sea-creatures-light-ocean-floor). Deheyn, *et al.* (1999) reported that the luminescence is about 100 times higher in intertidal luminous animals than in subtidal luminous animals. Further, in intertidal animals, the light emission is said to be two times higher during day than at night. Martini *et al.* (2019) reported that the light emission for benthic animals occurred at longer wavelengths (460–520 nm) than that of mesopelagic organisms (440–500 nm).

Functions of Luminescence

Much uncertainty still exists about the possible functions of luminescence in marine animals. Many deep-sea forms living in complete darkness use the bioluminescence as the only source of light. The light so produced helps animals with eyes to locate or recognize individuals of their own species or their prey. Thus luminescence serves as an important source of light in the dark depths of the ocean. Further, bioluminescence is believed to be helpful to attract the passing prey or to meet the opposite sex. In the case of fishes, it helps to keep them together in large shoals. Bioluminescence may also serve as recognition signs, a means of communication or courtship display in some organisms. For example both sexes of the polychaete *Odontosyllis* produce light during spawning. Similarly, the luminescent organs on the tips of barbels and anterior fin rays of stomiatoid fishes such as *Eustomias* and *Chirostomias* and *Ceratias* and other ceratioid angler-fishes, function as a lure for attracting the prey. In some deep sea shrimp *Systellapsis,* squid *Heteroteuthis,* mysid *Gnathophausia* and teleost *Malacocephalus* which discharge a luminous cloud into the surrounding water when irritated, light production serves as an aid to confuse or frighten their predators (http://ecoursesonline.iasri.res.in/mod/page/view.php?id=86809)

Chapter 2

Bioluminescence in Marine Invertebrates

Bioluminescence is seen in many marine organisms: which include bacteria, algae, jellyfish, worms, crustaceans, sea stars, fish, and sharks to name just a few. The number of species that luminesce and the variations in the chemical reactions which produce light suggest that bioluminescence has evolved many times say at least 40 separate times. This number however, continues to grow as the present research efforts make new discoveries.

Diversity of Light-Producing Marine Organisms

In the marine environment, the pelagic ecosystems support a significantly higher percentage of bioluminescent animals compared to the benthic ecosystems (76 per cent vs 32 per cent) (Martini *et al.*, 2019). The partial list of bioluminescent marine organisms hither-to discovered is given below:

1. Bacteria
2. Fungi
3. Phytoplankton : Dinoflagellates
4. Protozoans : Radiolarians
5. Cnidaria (= Coelenterata)
   - 5.1. Hydrozoans : Hydroids, Hydromedusae and Siphonophores
   - 5.2. Scyphozoans : True Jellyfish
   - 5.3. Anthozoans : Sea Pansies and Sea Pens
6. Ctenophores (= Acnidarians) : Comb Jellies
7. Nemertean worms : Interstitial ribbon worm

8. Annelids : Polychaete worms
9. Pycnogonids : Sea spiders
10. Crustaceans
    10.1. Amphipods
    10.2. Copepods
    10.3. Ostracods
    10.4. Decapod shrimps
    10.5. Euphausiid krills
11. Chaetognaths : Arrow-worms
12. Molluscs
    12.1. Bivalves : Clams
    12.2. Gastropods : Nudibranchs and Snails
    12.3. Cephalopods : Squid and octopods
13. Echinoderms
    13.1 Sea stars
    13.2. Brittle stars
    13.3. Crinoid
    13.4. Sea cucumbers
14. Hemichordate worms
15. Urochordates
    15.1. Pyrosomes
    15.2. Ascidian
    15.3. Larvaceans
16. Chordates
    16.1. Sharks
    16.2. Fish

Source : Haddock, http : //photobiology.info/Haddock.html; https://biolum.eemb.ucsb.edu/organism/)

Bioluminescent Capability of Marine Taxonomic Groups of Pelagic and Benthic Ecosystems

The pelagic ecosystems have been reported to possess higher taxa of marine taxonomic groups compared to the benthic ecosystems. The percentages of bioluminescent species in the different taxonomic groups of the pelagic and benthic ecosystems are given below:

Pelagic Ecosystems

Hydromedusae	100 per cent
Scyphozoa	96.6 per cent
Appendicularia	93.8 per cent
Annelida	> 90 per cent
Ctenophora	> 90 per cent
Siphonophora	> 90 per cent
Crustacea	68.2 per cent
Cephalopoda	51.8 per cent
Fishes	46.6 per cent
Chaetognatha	11.2 per cent
Pteropoda	6.4 per cent
Thaliacea	2.5 per cent
Benthic Ecosystems	
Octocorallia	100 per cent
Holothuroidea	94.3 per cent

Source: Martini et al., 2019

Depth-wise Occurrence of Bioluminescent Taxa

In the photic zone of about 100 m depth, the bioluminescent taxa relating to Hydromedusae, Siphonophora, and Ctenophora, are dominant. On the other hand, Chaetognatha are found to be dominant almost continuously, from 0 to 3,900 m depth while Thaliacea are mainly present shallower than 2,100 m. Among the decapod shrimp, 41 per cent occurring between 500 and 1,000 m depth are believed to be bioluminescent and the mesopelagic shrimp *viz.* euphausiid krill is mainly observed above 500 m depth (Martini and Haddock,2017)

Spectral Properties of Bioluminescence in Nektonic and Benthic Marine Invertebrates

Blue light with shorter wavelengths (400-500 nm) and green light with fairly longer wavelengths (500-600 nm) travel in waters deeper than 100 m. While the coastal organisms produce green light (490-520 nm), the vast majority of pelagic and deep sea bioluminescent organisms display a blue light with wavelengths of emission maxima (λmax) ranging from 450 to 490 nm (Gouveneaux *et al.*, 2017). Further, unlike blue or green light, the red light with wavelengths of emission maxima 600-700 nm is absorbed quickly. That is this red light fails to reach the deep-sea zone while it is travelling on the shallower sea, The spectral properties in terms of wavelengths of emission maxima of certain marine bioluminescent invertebrates are given below.

Emission Maxima (λmax) of Marine Invertebrate Animals

Animals	*λmax (nm) range*
Anthozoans	444-488
Annelid worms	565
Crustaceans	444-492
Molluscs	449-514

Source: Latz *et al.*, 1988.

Chemistry of Bioluminescence

Marine invertebrate bioluminescent systems which are using imidazopyrazinone compounds (*e.g.* luciferin types) as the reaction substrates are divided into luciferase and photoprotein types. The bioluminescent reactions involving luciferases are typical enzymatic reactions in which the substrate luciferin is oxidized by oxygen resulting oxyluciferin in the excited state whose relaxation to the original ground state is accompanied by the light emission. This type of reaction is seen in the luminescent invertebrates such as crustaceans (ostracods and deep-sea shrimp *Oplophorus*) and in anthozoan corals (*Renilla* spp.). On the other hand, in the bioluminescent systems of photoprotein type, which is seen in hydromedusae and ctenophores, the protein forms a stable enzyme substrate complex which is existing for a long time. This Ca+-regulated protein is a complex of protein with coelenterazine activated by oxygen, 2-hydroperoxycoelenterazine. A bioluminescent reaction is initiated in response to binding of calcium ions with Ca2+ binding sites on the surface of the protein molecule. Binding of Ca2+ causes small conformational changes in a substrate-binding cavity of the protein, which disturbs the hydrogen bind network that stabilize 2-hydroperoxycoelenterazine, thereby triggering the reaction of oxidative decarboxylation leading to the formation of the product in the excited state. Since a substrate whose oxidation is necessary for light emission is already bound in photoproteins, these, unlike luciferases, can react only once. *i.e.* it cannot turn over several times like a usual enzyme. It should be noted that since the reaction involves one molecule, the amount of the emitted light is always in proportion to the amount of the protein (Markova and Vysotski,2015)

Bioluminescence Reactions

Markova and Vysotski (2015) reported on the luciferin types and their distribution in bioluminescent marine invertebrates. The general scheme of bioluminescent reactions of luciferase and photoprotein types is given below:

Luciferase Type

Luciferase. +. O2 → Luciferase – Oxyluciferin. → Light (455-480 nm)

+ -CO2

Luciferin

Photoprotein type (Ca2 + -regulated photoprotein)

+3 Ca2+

Photoprotein → [Ca2+ -discharged photoprotein] → Light (455-480 nm)

- CO2

Bioluminescent Reactions of Luciferase and Photoprotein Types

Though there are about 30 known bioluminescent systems, the luciferin–luciferase pairs of only 11 systems have so far been characterised. Further, of these 11 systems, d-luciferin, coelenterazine and the bacterial bioluminescent systems have been fairly well understood and they are also found to be useful in biotechnological applications (Syed and Anderson, 2021). As the d-luciferin belongs to the bioluminescent insects, the remaining two are commonly found in bioluminescent marine invertebrates. The role of coelenterazine and bacterial luciferin in the bioluminescent systems is given below:

Coelenterazine

Coelenterazine which is the other most commonly used luciferin is seen in several marine taxonomic groups including copepods and the decapod deep-water shrimp. Coelenterazine is the substrate for about 15 different naturally occurring luciferases including the anthozoan *Renilla* luciferase (Rluc), copepod, *Gaussia* luciferase (Gluc) and copepod *Metridia longa* luciferase (Mluc). The bioluminescent reaction of coelenterazine also involves an enzymatically catalysed oxidation. Coelenterazine is converted to an excited state coelenteramide oxyluciferin through a dioxetanone intermediate. The excited state oxyluciferin relaxes to its ground state to emit a photon of blue light of wavelength 454–493 nm, dependent upon the enzyme. Further, this bioluminescent reaction is not dependent on ATP.

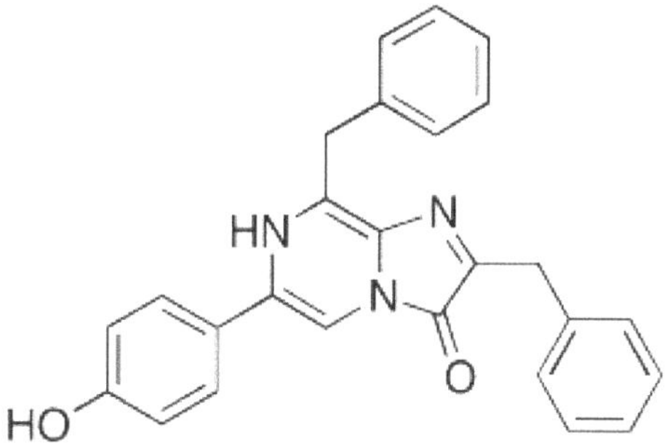

HN N N O HO

Coelenterazine

Vargulin

The vargulin (= *Cypridina*-type luciferin) is seen in the seed shrimp *i.e.* ostracods *viz. Vargula* and *Cypridina* which are known to synthesize this molecule from the amino acids arginine, isoleucine, and tryptophan (Vargulin. https://en.wikipedia.org/wiki/Vargulin).

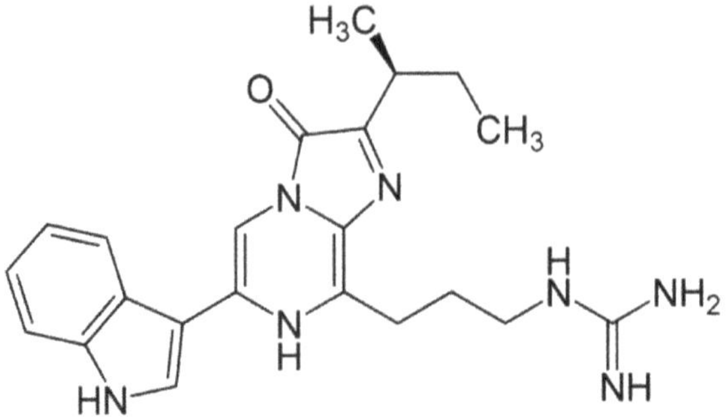

Vargulin

Bacterial Luciferin

All known bioluminescent bacteria are Gram-negative, facultative anaerobic and most of them are symbiotic and found in their host animals such as the Hawaiian Bobtail squid (*Aliivibrio fischeri*). Some are however, like *Vibrio harveyi* and *Alteromonas hanedai* Bacterial luciferase which is a flavin-dependent monooxygenase is remarkable for its characteristic feature in transforming chemical energy to photons of visible light. This bacterial luciferase catalyzes bioluminescent reaction using reduced flavin mononucleotide (as a luciferin), long-chain aldehyde and oxygen to yield oxidized flavin, corresponding acid, water and light at λmax around 490 nm (Tinkul *et al.*, 2020).

Other Marine Luciferins/Luciferases

Luciferin	*Luciferase Size*	*Wavelength (Emission max)*
Cypridina luciferin(CLuc)	~ 61 kDa	452 nm; 465nm
Fungal luciferin	~ 29 kDa	520 nm
Dinoflagellate luciferin	~ 135 kDa	476 nm
Krill luciferin	~ 600 kDa	468 nm
Odontosyllis luciferin	~ 35 kDa	510 nm
Sea pancy luciferin (Rluc) (coelenterazine)	~ 36 kDa	480 nm
Copepods (GLuc) (coelenterazine)	~ 20 kDa	480 nm
Copepods (MetLuc) (coelenterazine)	~ 24 kDa	480 nm

Source: Syed and Anderson, 2021.

Luciferin Type and Coelenterazine in Luminescent Marine Invertebrates

In the marine luminescent animals, the coelenterazine and vargulin are the major types of luciferin in the luminous marine invertebrates. Further, among these two luciferins, colelenterazine has been reported to be present in more number of species in the crustacean deep-sea shrimps (Hoppe *et al.*, 2012; Haddock *et al.*, 2010) as shown in following tables

Luciferin Type in Luminescent Marine Invertebrates

Phylum/Group	*Luciferin Type*
Cnidaria - Hydrozoa and Anthozoa	Coelenterazine
Ctenophora	Coelenterazine
Nemertea	Unknown
Annelida -Polychaeta	Unknown
Amphipoda	Unknown
Pycnogonida	Unknown
Crustacea- Ostracoda	Coelenterazine and Vargulin
Crustacea -Mysid shrimp	Coelenterazine
Crustacea - Decapod shrimp	Coelenterazine
Mollusca- Bivalvia	Other -known
Mollusca- Gastropoda	Unknown
Mollusca- Cephalopoda	Bacterial (+ symbiont); Coelenterazine;unknown
Echinodermata- Asteroidea	Unknown
Echinodermata-Ophiuroidea	Coelenterazine; unknown
Echinodermata-Crinoidea	Unknown
Echinodermata-Holothuroidea	Unknown

Distribution of Coelenterazine Type Luciferin in Marine Invertebrates

Phylum/Class	*Genera/Species*
Cnidaria – Hydrozoa	*Aequorea, Obelia*
Cnidaria - Anthozoa	*Acanthoptilum, Cavernularia, Ptilosarcus, Renilla,Stylatula*
Ctenophora	*Beroe, Mnemiopsis*
Crustacea- Mysidacea	*Gnathophausia ingens*
Crustacea – Decapoda (shrimp)	*Acanthophyra eximia, Acanthophyra purpurea, Heterocarpus grimaldii, Heterocarpus laevigatus, Oplophorus spinosus, Systellapsis cristata, Systellapsis debilis*
Mollusca - Cephalopoda	Watesenia (squid)
Echinodermata-Ophiuroidea	*Amphiura filiformis*

Chapter 3

Biology and Ecology of Bioluminescent Marine Invertebrates

3.1. Marine Sponges

Phylum: Porifera; Class: Demospongiae; Order: Suberitida; Family: Suberitidae

Suberites domuncula

Image credit: Guido Picchetti, Commons Wikipedia

Common name (s): Sea-orange

Global distribution: Subtropical to polar: Arctic, Northeast Atlantic and the Mediterranean

Habitat: This sessile species gets fixed mainly on the shells inhabited by the hermit crab, *Paguristes eremita* at a depth range of 0 - 1100 m.

Biology: This sponge is more or less spherical with a maximum dia. of 10 cm. It has smooth surface with only some small channels which may often even one and single. Coloration of body is variable from orange to marbled light blue.

Bioluminescence: The biosensor system of this luminous sponge employs its light producing (organic) luciferase-like proteins, (inorganic) light transducing silica spicular system and a cryptochrome as the source of light, optical waveguide and photosensor, respectively. The light transmission features of its smaller spicules (200 µm) and the ability of the sponge tissue to produce light have been well demonstrated (Wang *et al.*, 2012).

Martini *et al.* (2020) however, reported that this marine sponge uses a firefly luciferin and firefly luciferase homolog as its luminescence system. Because these sponges filter large volumes of water, it is very difficult to distinguish between luminescence of the animal itself or light produced by other organisms present within its tissues. It is also suggested that the bioluminescence in sponges may have been induced by the bioluminescent symbiotic or captured bacteria living in these sponges.

Order: Poecilosclerida Family: Cladorhizidae

Unidentified species (*Cladorhiza* sp. ?)

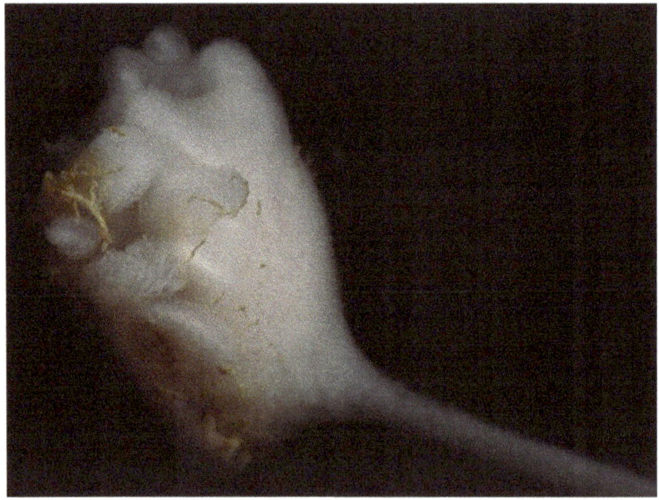

Image credit: Martini, S., Schultz, D.T., Laundersen, L. and Haddock, S.H.D. Applied for permission

Common name (s): Not applicable

Global distribution: Northeast Pacific Ocean

Habitat: This deep-sea sponge was collected from seafloor composed of silt and clay.

Biology: This sponge which was anchored to the substrate with rhizoids resembled a globular mass,length and width of which ranged from 8 to 18 mm and 5to 14 mm respectively. It showed numerous filamentous processes of 9–15 mm length. Spicule observations showed unguiferous anisochelae of 55–59 µm size.

Bioluminescence: This sponge has been reported to emit blue-green light after gentle touching with a gloved hand or round-pointed forceps. Further, the luminescence in the animal occurred in various parts of the animal rather than in a single location. The light kinetics of this sponge were bright and visible even to the naked eye for 5–10 s. Further, repetitive stimulations were reproducible, and the light did not appreciably dim over time (Martini, *et al.*, 2020).

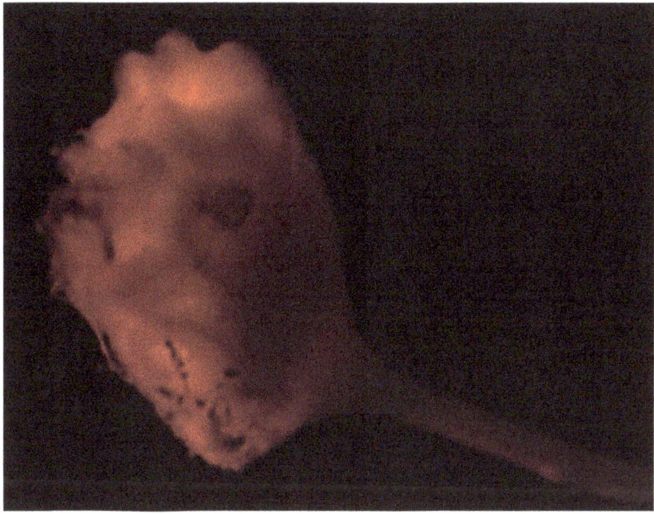

Image credit: Martini, S., Schultz, D.T., Laundersen, L., Haddock, S.H.D. Applied for permission

3.2. Cnidarians

Among the different classes of coelenterates *viz.* Hydrozoa (the jellyfish, Aequorea); Scyphozoa (the jellyfish, *Pelagia*); and Anthozoa (the sea cactus, *Cavernularia*; sea pansy, *Renilla*; and sea pen, *Leioptilus*), there is a significant uniformity in the substances involved in light-emitting reactions *i.e.*, "luciferins" and photo-protein chromophores, as well as the light-emitter product. In each of these instances, the reaction product namely, 2-(p-hydroxy-pnenylacetyl) amino-3-benzyl-5-(p-hydroxyphenyl) pyrazine is the actual light-emitter, whether it occurs in a Ca2+-triggered photoprotein type of luminescence, or in a "luciferin-luciferase" type. In certain coelenterates, *e.g.*, *Cavernularia*, the above two types are equally significant; in others (*Renilla* and *Leioptilus*),the "luciferin-luciferase" type predominates over the Ca-triggerable photoprotein type; and in the luciferaseless jellyfish, Aequorea, only the photoprotein type functions. However, in all instances studied, the structure of the light-emitter prior to the luminescence reaction appears to be essentially the same (Shimomura and Johnson. 1975).

In benthic ecosystems of the deep-sea, octocorals are some of the most abundant luminous animals. Among such luminous sessile animals, the shallow-water sea pansy *Renilla* has been well studied for its chemistry. However, aside from *Renill*a, little is known about the bioluminescent mechanisms of other anthozoans, especially deep-sea octocorals. Recent studies on the bioluminescence of the deep-sea anthozoans, coelenterazine-dependent luciferase activity was found to be dominant. Bessho Ueharan *et al.* (2020) reported that all luminous octocorals may share a common biochemical mechanism, which utilizes coelenterazine and *Renilla*-type luciferase.

Phylum: Cnidaria; Class: Anthozoa; Order: Actiniaria; Family: Actinoscyphiidae

***Actinoscyphia* sp.**

Image credit: Aquapix and Expedition to the Deep Slope 2007, Common Wikipedia

Common name (s): Venus flytrap sea anemone

Global distribution: Gulf of Mexico, West Africa and American Samoan region of the Pacific,

Habitat: It is found attached to the stalk of a dead glass sponge in muddy situations at bathyal depths (about 650 m) in deep water canyons and upwelling region

Biology: Its pedal disc is small, and its tentacles are short compared to the large, concave oral disc, which is funnel or mushroom-shaped. It extends its tentacles in two rows, one reflexed back and one sloping forward.

Bioluminescence: The bioluminescence of this unidentified species consisted of adhesive luminescent secretions (or possibly detached luminescent epidermis). Its emission peak wavelengths were reported as 483nm and 455 nm (Johnsen *et al.*, 2012).

Order: Malacalcyonacea; Family: Epizoanthidae

Epizoanthus induratum

Image credit: iNaturalist; Wikipedia

Common name (s): Luminescent zoanthid

Global distribution: It has cosmopolitan distribution.

Habitat: This sessile species occurs at a depth of 75 m

Biology: It is a bilaterally symmetrical and soft bodied animal. The species of this family which are defined by macrocnemic mesenterial arrangement and single mesogleal marginal sphincter muscle lack a canal system and lacunae in the mesoglea of the column. Cell islets are present.

Bioluminescence: In this luminescent species, its wavelength of emission maximum has been reported to be 500 nm (Widder *et al.*, 1983).

Family: Paramuriceidae

Acanthogorgia sp.

Image credit: NOAA/Monterey Bay Aquarium Research Institute, Commons Wikipedia

Common name (s): Gorgonian sea fan, gorgonian coral

Global distribution: Similan Islands, Andaman Sea, Thailand

Habitat: This species is found distributed on the slopes of the rocky substratum between depths of 14 and 24 m.

Biology: Colony of the species of this genus is flabellate or bush-like with slender branches. Coenenchyme layer of branches is very thin. Polyps are monomorphic, non-retractile, cylindrical, and are taller than wide. Warty spindles are found arranged in eight chevron rows along polyp walls. Spindles are straight, bent, and are sometimes bifurcated. Tentacles are covered with small, flat sclerites.

Bioluminescence: It is a bioluminescent species (Raddatz *et al.*, 2011).

Order: Scleralcyonacea; Family: Chrysogorgiidae

Chrysogorgia desbonni

Common name (s): Not known

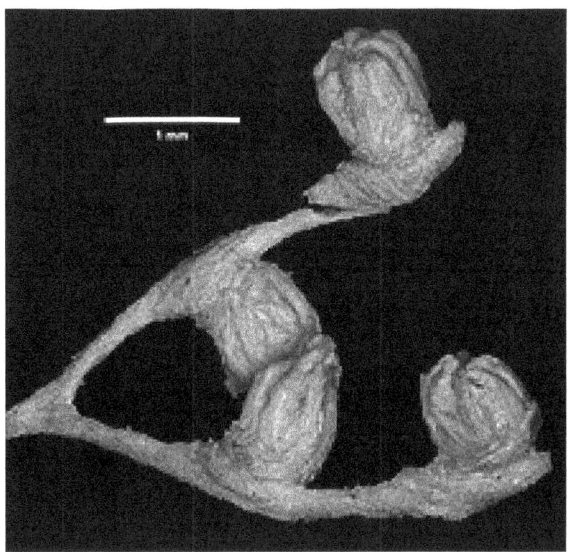

Image credit: Daniel J. Drew - Gall L, Wikispecies

Global distribution: North Atlantic Ocean; Caribbean Sea

Habitat: It occurs on seamounts at depths of 315 – 330m.

Biology: Branching of the colonies of the species of this genus is sympodial, in the form of an ascending spiral (clockwise or counterclockwise). A fan (planar colony), or two fans are emerging from a short main stem.Axis has a metallic shine and dark to golden in color. Branches divide dichotomously or pinnately. Most polyps are large relative to the size of the branches Sclerites are in the form of spindles, rods and scales are with little ornamentation.

Bioluminescence: It is a species of bioluminescent coral (Johnsen *et al.*, 2012)

Family: Chrysogorgiidae

Iridogorgia sp.

Common name (s): Not known

Global distribution: North Atlantic Ocean

Habitat: It occurs in deep-sea coral habitats.

Biology: In the species of this genus, the main axis is monopodial and is spiraling upward, with undivided branches emanating from one side of axis. In sexually matured colonies, polyps are short with base expanded along upper side of branch. Sclerites rods, spindles, or scales are arranged vertically, extending in tracts onto tentacles. Branch coenenchyme is with sclerites oriented along branch or without sclerites between polyps.

Bioluminescence: It is a bioluminescent species (Raddatz *et al.*, 2011)

Image credit: NOAA, Wikimedia Commons

Family: Corallidae

Heteropolypus ritteri

Image credit: NOAA,Wikimedia Commons

Common name (s): Golf Ball coral, mushroom soft coral

Global distribution: Eastern Pacific; off the coast of California and Mexico

Habitat: This benthic species occurs on hard substrate up to depths as great as 3,000m.

Biology: This soft coral resembles the stony corals, but lack the distinctive stony skeleton. Polyps are with only eight showy tentacles which contain poisonous stinging cells. It gets its name because when the polyps are pulled in, it looks like a mushroom.

Bioluminescence: The autozooids and siphonozooids of this species have been reported to emit blue light when the animal was stimulated with KCl. Further the light emission was slow and dim. The peak wavelength of the bioluminescent spectrum was found to be 476 nm (Bessho-Uehara *et al.*, 2020).

***Anthomastus* sp.**

Image credit: Deepwater Coral Expedition, Wikimedia Commons

Common name (s): Deep-sea mushroom coral

Global distribution: North Pacific Ocean

Habitat: This benthic species occurs at depths between 280 and 330 m.

Biology: The colony of this species is very small in size, measuring 15 mm in total height.Foot is very short and is 2 to 3 mm in high. Its globular capitulum of 16mm in dia. bears 16 autozoids which are regularlyscattered on its surface. Zone of separation of the capitulum and the foot is only little marked.Siphonozoids are not visible to the naked eye, but are observable on a tangential section at the surface of the capitulum. Autozoids are more or less contracted and, most often they are alone; and the tentacles are expanded and laterally flattened.

Bioluminescence: Luminescence has been observed in this unidentified species (Raddatz *et al.*, 2011).

Paragorgia arborea

Image credit: NOAA/Monterey Bay Aquarium Research Institute; Wikimedia Commons

Common name (s): Bubblegum coral

Global distribution: It is found widespread in the Northern Atlantic and Northern Pacific Oceans

Habitat: It occurs on seamounts and knolls at depths of 200- 1,300 m.

It is often found associated with the Gorgon's Head basket star *Gorgonocephalus caputmedusae.* It sometimes forms dense coral gardens with other octocorals, such as *Primnoa resedaeformis, Paramuricea grandis* and *Keratoisis ornata* and the sea pen *Pennatula borealis.*

Biology: It has fan-shaped structure with a tough central trunk and many branches. Branch tips are bulbous, giving this species its common name, bubblegum coral. It has an average growth rate of 1cm/yr. It is brightly colored in white or red and it can grow to heights of 6 m.

Bioluminescence: It has been reported to emit blue light upon mechanical stimulation (Bessho-Uehara *et al.*, 2020).

Family: Echinoptilidae

Actinoptilum molle

Common name (s): Radial sea pen or purple sea pen

Global distribution: South Africa

Habitat: This coastal species is generally found in sand or silt bottoms and sometimes in sandy areas on rock substrate at depths of 12-333 m

Biology: Colonies of this species range in length up to about 240mm. They have a symmetrical slightly tapering round-tipped cylindrical rachis and a tapering peduncle, the size of which is between one fifth and one third of the total length of the colony. Rachis is covered all round with dimorphic polyps which are radially arranged with respect to the longitudinal axis. Siphonozoids are found packed between the bases of the retractile autozooids, which possess inconspicuous non-retractile bifurcated calyces. Color of the colony is variable and individual colonies may be reddish brown, pink, yellow, white or cream; rachis may be purple to reddish purple; and peduncle is yellow, white, pink or brownish.

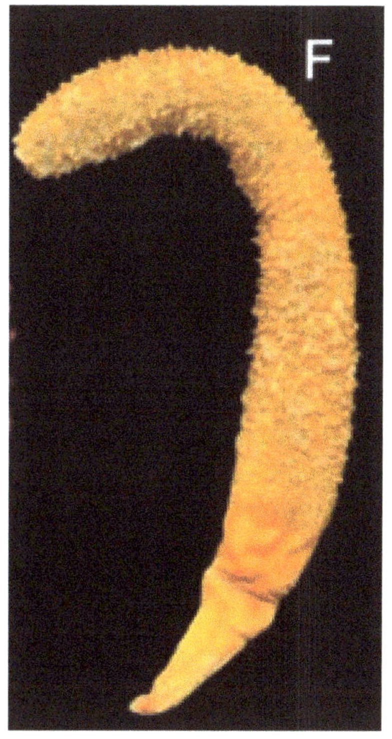

Image credit: Gary Williams, Reproduced with permission

Bioluminescence: It has been reported to emit green light when agitated (Francis and Vilar, 2020).

Family: Funiculinidae

Funiculina sp.

Common name (s): Tall sea pen

Global distribution: North Atlantic and Mediterranean; New Zealand

Habitat: This sublittoral to deep offshore water species occurs in deep muds, especially in sea lochs at depths of 20-2000m.

Biology: It is a tall, narrow sea pen, which can exceed 2 m. in height. It has a calcareous white axis which is square in section. Lower quarter of stem forms a smooth peduncle and upper one third is curved. Autozooids which are often pink get irregularly arranged on rachis or in short oblique rows.

Bioluminescence: The species of this genus have been reported to emit blue light upon mechanical stimulation of the stalk or polyps. Gentle squeezing of polyps could trigger light emission from the stalk. The wavelength of maximum emission was 485 nm. Further, the addition of CaCl2 and coelenterazine was found to be sufficient to yield light from a 1 cm long piece of the stalk (Bessho-Uehara *et al.*, 2020).

Family: Halipteridae

***Balticina* sp** (= *Halipteris* sp.)

Common name (s): Sea pen

Global distribution: Bering Sea., San Antonio Cape (Buenos Aires coast)

Habitat: This benthic species occurs at depths of 440-480m

Biology: Unlike the species of *Renilla*, the species of this genus are slender and rod-like.

Bioluminescence: In this luminescent species, its emission peak wavelength was 510 nm and it pulsed with a period of ~0.5s (Johnsen *et al.*, 2012)

Image credit: Hawaii Undersea Research Laboratory Archive. Applied for permission

Family: Keratoisididae

Acanella sp.

Image credit: NOAA, Wikimedia Commons

Common name (s): Not known

Global distribution: North Atlantic Ocean, Gulf of Mexico, Mediterranean Sea, central Pacific Ocean, Indo-Pacific, and Indian Ocean

Habitat: This benthic species occurs at depths of 300–2121 m

Biology: Colony of this species is bushy. Sometimes it is with a distinct main stem, from one to about six branches per node. Internodes are solid, smooth or longitudinally striated. Polyps are covered with finely thorned or warted spindles, sometimes with rods in the tentacles. Coenenchyme is thin, with irregularly distributed spindles. Its large fusiform spicules (=sclerites) are obliquely arranged up the polyp body and the eight long spine-like spicules projecting between each tentacle.

Bioluminescence: In this luminescent species, its different polyps turned on and off asynchronously in a 'twinkling' display. Its emission peak wavelength was 480 nm (Johnsen *et al.*, 2012).

Isidella tentaculum

Common name (s): Bamboo coral

Global distribution: It has worldwide distribution except for Arctic waters

Habitat: It is an inhabitant of the deep sea; and occurs in rocky seafloor, including seamounts and submarine canyons at depths of 400–4,850 m.

Image credit: NOAA Office of Ocean Exploration, Wikimedia Commons

Biology: The stony branches of this species contain thousands of tiny polyps living and working together. The individual polyps ae known to stretch their feathery tentacles into the currents to grasp plankton and other particles of food drifting in the currents. The skeletons of bamboo coral are made up of calcium carbonate the joint-like nodes or axes are composed of gorgonin protein. The life span of a typical bamboo coral has been estimated to be between 75 and 126 years.

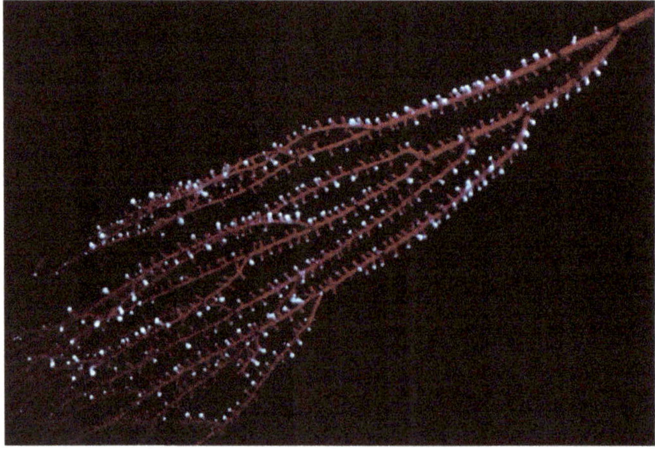

Image credit: Sonke Johnsen
(Reproduced with permission)

Bioluminescence: This species has been reported to emit bright blue light and secrete luminous mucus upon strong stimulation. However, emission of green -blue light has also been observed in this species upon agitation with 483 nm as the wavelength of emission maximum (Bessho-Uehara *et al.*, 2020).

Keratoisis flexibilis

Image credit: WoRMS (It does not exercise any editorial control).

Common name (s): Not known

Global distribution: Eastern Atlantic and Caribbean Sea

Habitat: This sessile species is found on the outer continental shelf and slope at a depth range of 170 - 592 m.

Biology: In this species, the branching is in one plane, but it may take an irregular appearance as and when its quite long and flexible branches tend to curl. It arises from a basal calcareous disc with the lengths of the internodes successively increasing from the base and the longest internode is 17 mm. Internodes are hollow and are coarsely and regularly grooved in the medial and distal parts of the colony. Polyps are squat and are generally well spaced and are rarely opposite.

Bioluminescence: This species has been reported to bioluminesce with a blue glow when it is touched (https://oceanexplorer.noaa.gov/explorations/05deepcorals/logs/nov12/media/movies/bioluminescing_bamboo_video.html)

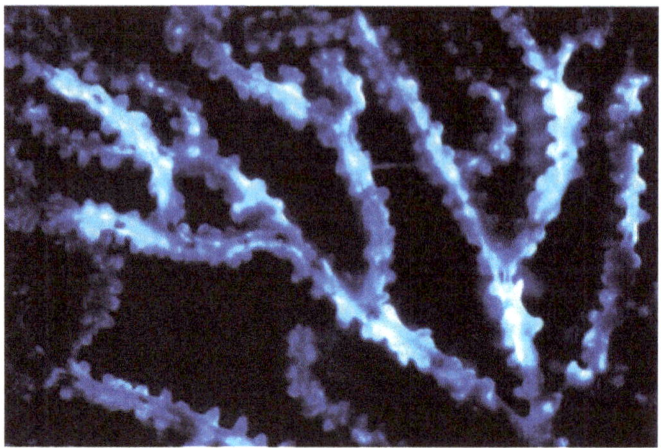

Image credit: Edith Widder. (Applied for permission)

***Keratoisis* sp.**

Image credit: Wikimedia Commons

Bioluminescence: Raddatz *et al.* (2011) reported that a strong luminescence was accidentally noticed when this unidentified bamboo coral species was brought to the deck in the early evening hours just before sunset. Initially, the entire stem and branch tissue displayed a dull blue luminescence. Further, when it was touched, it emitted a very strong blue light that lasted for a few seconds. Its tissue lit up strongest and flash like at the point of stimulation, and the illumination spread in a wave across the coenenchyme of the distal branches. The most intense light emission however, originated from its non-retractile scleriterich feeding polyps and this display remained visible for several minutes before it slowly faded. This phenomenon could be noticed several times within hours

Lepidisis olapa

Common name (s): Bamboo coral

Global distribution: Eastern Central Pacific: Hawaii.

Habitat: This sessile species is *found in the deep sea* at a depth of about 700m.

Biology: This species is branched from the nodes or the internodes which are hollow. Polyps are erect and their walls are containing spindles and needles, eight of which project beyond the margin. Their surface is covered with small scales. Tentacles are also covered with scales. Coenenchyme is also filled with scales similar to those of the polyps.

Bioluminescence: This unbranched is bioluminescent and is creating blue light when touched. It is also reported that the light shoots up and down the axis (https://oceanexplorer.noaa.gov/explorations/04alaska/background/bamboo/bamboo.html)

***Lepidisis* sp.**

Image credit: NOAA, Wikimedia Commons

Muzik (1978) reported that during submarine dives, the living colonies of this deep sea species were luminescent when they were mechanically stimulated. Johnsen *et al.* (2012) reported that its emission peak wavelength was 475 nm.

Bessho Uehara *et al.* (2020) made a detailed study on the luminescence of this species both in filed and in lab. According to them, its blue-light emission propagated as a wave and the light intensity was particularly strong on the polyps. However, the whole body including along the stem was found capable of light emission. In the lab, when one of the polyps was stimulated, blue flashes were emitted from the tips of polyps and the light was distributed as a wave, from the point disturbed to the tip of the next polyp. When it was chemically stimulated with KCl, the whole animal glowed continuously for a few minutes and the specimen had green fuorescence localized along the stem and in polyps under UV excitation. At this time the wavelength of emission maximum was recorded as 365 nm. Interestingly, a few polyps emitted pink fuorescence and the peak wavelength of the bioluminescent spectrum was at 477 nm. Further, light was produced upon the injection of the coelenterazine, suggesting the presence of coelenterazine-dependent luciferase. Furthermore, upon injection by coelenterazine, the light intensity of the luciferase sample rose significantly and the luciferase activity was higher at pH 7.5 than pH 8.6. The luciferase activity was also sensitive to divalent cations of Ca2+ and Mg2+ at 50 mM. These results suggested that the bioluminescence of *Lepidisis* is produced by coelenterazine and a coelenterazine-dependent luciferase system

Family: Kophobelemnidae

Kophobelemnon stelliferum

Image credit: Kåre Telnes, Reproduced with permission

Common name (s): Wild star sea pen

Global distribution: Atlantic Ocean, Pacific Ocean and the Mediterranean.

Habitat: This sessile species is known from circalittoral (The region of the sublittoral zone which extends from the lower limit of the infralittoral to the maximum depth at which photosynthesis is still possible) and bathyal zones. It has also been reported to thrive on muddy substrate, on depths ranging from 25 m. to 1600 m.

Biology: The polyps of this species grow directly from the stalk. The white or pale brownish or sometimes purple polyps are generally larger than on the tall sea pen. The colony can grow to a length of 75 cm, but most of the animal is found buried in the mud.

Bioluminescence: It is a bioluminescent species but the colour of light emitted by this species is not known (Francis and Vilar, 2020)

***Kophobelemnon* sp.**

Bioluminescence: Bessho-Uehara *et al.* (2020) reported that a specimen of this unidentified species collected at a depth of about 3980 m displayed green fluorescence at the bases between the tentacles. Further, during in situ observation this specimen upon agitation, emitted green light from the photophores present between the tentacles. The wavelength of its emission maximum was recorded as 538 nm.

Family: Mopseidae

***Primnoisis* sp.**

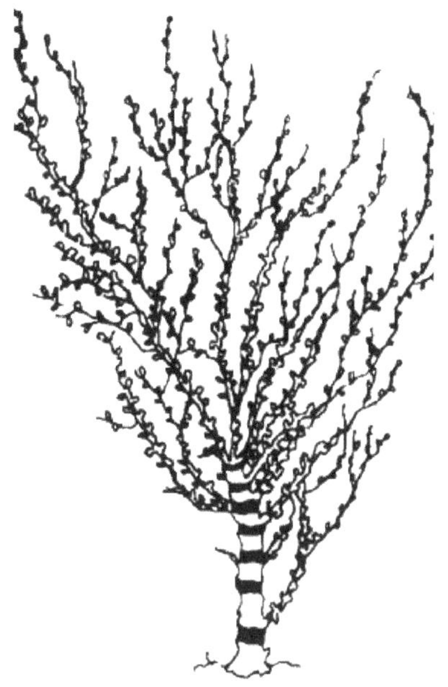

Image credit: Covadonga Orejas, Josep-Maria Gili, Wolf Arntz (Applied for permission)

Common name (s): Antarctic octocoral

Global distribution: Antarctic Ocean, Scotia Sea

Habitat: This deep sea species occurs at depths of 1228-1400m.

Biology: Not reported

Bioluminescence: This is a bioluminescent species (Raddatz *et al.*, 2011)

Family: Pennatulidae

Pennatula phosphorea

Common name (s): Phosphorescent sea pen

Global distribution: It is present widely in the north-east Atlantic and in the Mediterranean.

Habitat: It is found in soft, sandy and muddy substrata at depths of 10-100 m.

Biology: Colonies of this species are stout and fleshy and are up to 40 cm long. It is composed of a basal peduncle embedded in the substratum, from which arise leaf-like branches. Fused polyps form large triangular 'leaves' in alternate opposing lateral rows. Back of the peduncle is covered in inhalent polyps *viz.* siphonozooids. Flesh is translucent, yellowish or pale pink with white polyps. Colony may also be a deep reddish-pink due to the presence of red skeletal plates in its tissue. Central axis of the colony is often bent-over at the tip like a shepherds crook. Polyps are placed in one row within half a line of one another. Feeding polyps are however separate.

Image credit: MondoMarino.net, Wikipedia

Bioluminescence: This animal has been reported to luminesces when excited, and its blue-green light emanates from the zooids. Further, from the region stimulated, luminous waves procees to the extremities of the animal. Its transmission system is non-polarized because light waves can be initiated in either direction and up or down the colony(Nicol, 1958). Widder *et al.* (1983) reported its wavelength of emission maximum as 500 nm

Pennatula rubra

Common name (s): Red sea pen

Global distribution: Atlantic and the Mediterranean.

Habitat: This sessile species occurs on the sandy/muddy bottoms of the infra- and circumlittoral zones. It can move in the water using its polyp leaves as fins.

Biology: This species is in the shape of a feather or fern, and is reddish-orange coloured with a thick central white stripe along the rachis. Its polyps are placed in a double row and the animal's length may be around 7cm when fully contracted and 26cm when fully extended.

Bioluminescence: In this species, a soft bioluminescence was observed in an aquarium after electric stimulation but no evidence of its bioluminescent behaviour in nature has been published thus far (Chimienti, *et al.*, 2018). However, Francis and de Vilar (2020) reported that its green light was visible along the autozooids around the edge of the polyp leaves but sparsely along the rachis from the siphonozooids. Further, its light emission only occurred in dark-adapted specimens, but animals appeared to be luminous (or could be stimulated) regardless of time of day.

Image credit: Warren R Francis and Ana¨ýs Sire de Vilar (Applied for permission)

Pteroeides griseum (= *Pteroeides spinosum*)

Common name (s): Spiny sea-pen, gray sea pen, grey sea feather, armoured sea feather

Global distribution: It is quite rare and is found in the Mediterranean Sea

Habitat: It occurs on shifting bottoms from 30 to 250 m. deep.

Biology: It has a general shape of a feather, and bears numerous parallel polyp leaves on each side of the rachis. Unlike the species of *Pennatula*, siphonozooids appear to be absent along the rachis, and are instead concentrated at the base of the polyp leaves. It is approximately 6-10cm long when contracted and up to 32cm when fully extended.

Bioluminescence: In this species light emission occurs only in dark-adapted specimens, but animals appeared to be luminous (or could be stimulated) regardless of time of day. Green light was seen along the tips of the polyp leaf, and among a high density area of siphonozooids on the leaf (Francis and de Vilar, 2020).

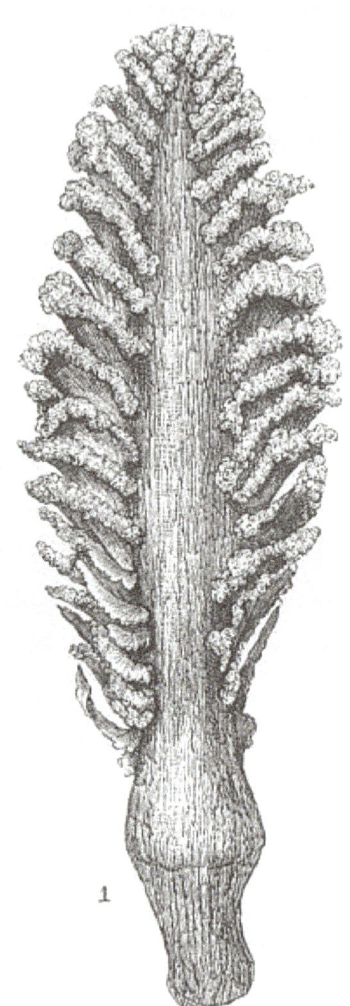

Image credit: Filhol, H., Wikimedia Commons

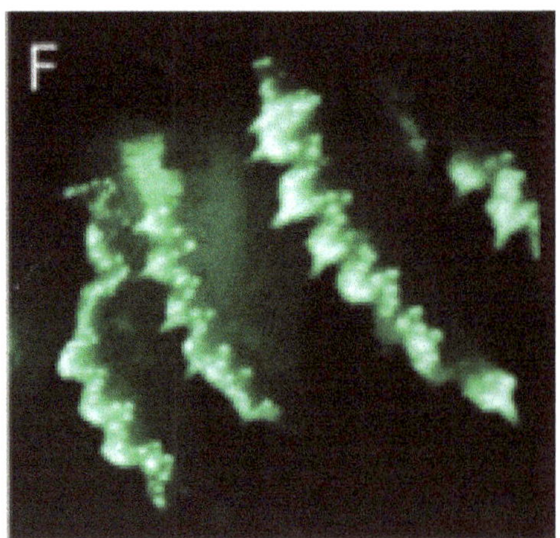

Image credit: W.Francis (Applied for permission)

Ptilosarcus gurneyi (= *Leioptilus gurneyi*)

Image credit: Stan Shebs, Wikimedia Commons

Common name (s): Orange sea pen or fleshy sea pen

Global distribution: It is native to the northeastern Pacific Ocean; western coast of North America and its range is extending from Alaska to southern California

Habitat: It lives in deep water anchored by its base in soft sand or mud substrates at depths ranging from 14 to 225 m

Biology: In this colonial cnidarian, the individual polyps possess their own specialized functions. The primary polyp, loses its tentacles and forms both the stalk of the colony (known as the rachis), and the bulbous base with which it anchors itself deep into the soft substrate. Other polyps which are known as secondary polyps include autozooids. These are feeding polyps, and are armed with cnidocytes on the eight branching tentacles which form the feathery branches of the sea pen. They also contain the gonads. Other secondary polyps which are siphonozooids, can force water into and out of the colony to ventilate it. When the colony is disturbed, it can pump water out and retract into its bulbous base. At this time it also emits bioluminescence in order to startle a potential predator. The animal resembles an old-fashioned quill pen. It grows to a height of 46 centime and its coloration may be white, yellow or orange

Bioluminescence: When contact is made by the leather star (the asteroid, Dermasterias imbricata)this animal has been reported to react by emitting a flash of light and dive into the sand (https://en.wikipedia.org/wiki/Ptilosarcus_gurneyi)

Davenport and Nicol (1956) reported that in this species, light is emitted by the polyps (autozooids and siphonozooids) when the animal is excited. Further, gentle mechanical stimulation and electrical stimulation evoke luminescent waves which are known to pass over the rachis at a velocity of 26·4 cm/s. During this exercise, the first luminescent wave appeared after 1 to 3 shocks, depending on the condition of the animal. Subsequently, light waves arose at the rate of 1/

shock, and increased progressively in intensity due to neuroeffector facilitation. On the other hand, strong mechanical and prolonged electrical stimulation produced a refractory state in which the animal contracted and there was no r production of luminescent waves. Slow progressive spread of luminescence was induced subsequently by maintaining repetitive stimulation.

***Ptilosarcus* sp.** (= *Leioptilus* sp.)

Bioluminescence: Shimomura and Johnson (1975) reported that the bioluminescence of this species involves both types of luminescence mechanisms i.e photoprotein and luciferin-luciferase types. Further, the luciferyl sulfate extracted from this unidentified species was chemiluminescent in dimethylsulfoxide or in dimethylformamide after acid-treatment of the samaples. Furthermore, when slices of this species were put into cold methanol, a fairly strong light emission was observed

Family: Primnoidae

***Thouarella* sp.**

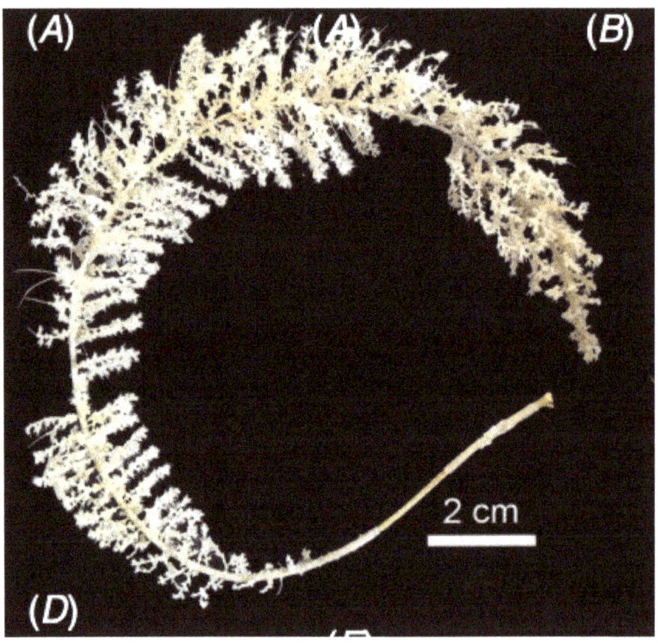

Image credit: Mónica Núñez-Flores A,B,C,D, Daniel Gomez-Uchida B and Pablo J. López-González. (Applied for permission).

Common name (s): Bottlebrush coral

Global distribution: Antarctic Ocean, Ross Sea; Antarctic and subantarctic waters.

Habitat: This deep sea species occurs at depths of 598 – 613m. It is known to provide habitat for a variety of animals such as polychaetes, anemones and crinoids

Biology: The species of *Thouarella* possess flower-like, open operculate polyps covered with thin sclerites. Colony consists of a main stem in which branching is either pinnate, dichotomous, or in a bottlebrush manner in which it is branched in at least 3 directions. Polyps are isolated, paired, or in whorls and are upwardly inclined at 45–90° from the branchlet. Polyp heads are wider than the base and are completely protected by 5-8 rows of longitudinally arranged scales and in each row, there may be 5–15 (generally 5–8) scales. However, adaxial body-wall scales are generally reduced in size and number.

Bioluminescence: It is a bioluminescent species (Raddatz *et al.*, 2011).

Family: Protoptilidae

Distichoptilum gracile (=*Distichoptilum verrilli*)

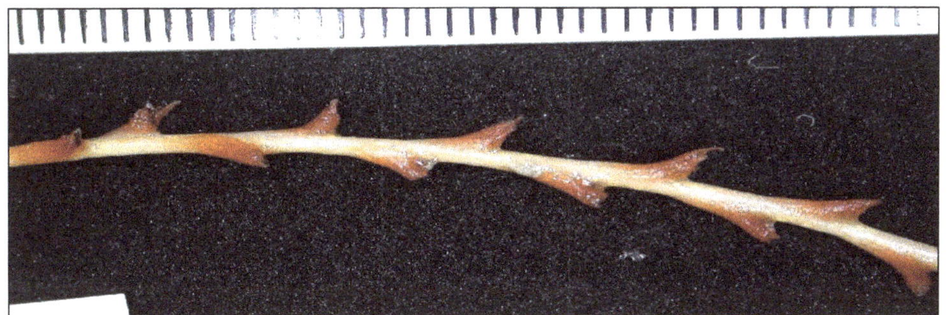

Image credit: CSIRO - Karen Gowlett-Holmes. (CC/NC)

Common name (s): Two-lined sea pen, saw-tooth sea pen

Global distribution: This species has an almost cosmopolitan but limited distribution in Atlantic, Indian and Pacific Oceans. In the Pacific Ocean, its occurrence is limited only to Mexico, the Galapagos Archipelago and off New Zealand.

Habitat: This deep-water pennatulacean is found in soft bottom at depths of 650–4300 m.

Biology: It is a long, very narrow sea pen with two oppositely-placed longitudinal rows of closely-adjacent polyp calyces along the rachis. The polyp calyces are elongate and acutely-pointed at the apexes. Colouration of the colony is a clear red-orange throughout.: Maximum length of the colony may be less than one metre.

Bioluminescence: The wavelengths of emission maximum of light from the stalk of this species were 505 nm and 508 nm (Johnsen *et al.*, 2012; Widder *et al.*, 1983)

Bessho-Uehara *et al.* (2020) reported that this species emitted green light waves upon mechanical stimulation. In the lab, when the polyps were touched, they emitted bright green light visible to the eye. Further, the addition of a few drops of KCl triggered bright sustained light which lasted for a few minutes. Furthermore, this species secreted luminous mucus apart from the appearance of green glow on the stalk. The wavelength of emission maximum was found to be 507 nm and the presence of a green fluorescent protein is also suggested in this species.

Protoptilum sp.

Common name (s): Worm-like sea pen

Global distribution: it is widely-distributed throughout much of the Pacific and northern Atlantic Oceans.

Habitat: This has been reported to occur at depths of 250–4000 m

Biology: It is an elongate, narrow sea pen with flattened polyp calyces arranged in oblique rows of less than 5 polyps per row. Calyces are variable and may have 0–8 terminal teeth. Colouration of the rachis is white to pale pink with red polyp calyces. Size of the colony is often less than 300 mm in length.

Bioluminescence: The emission peak wavelength of this luminescent species was 535 nm (Johnsen *et al.*, 2012)

Family: Renillidae

Renilla kollikeri

Image credit: Wikipedia Commons

Common name (s): Sea pansy

Global distribution: Southern coast of California, including Santa Barbara, California.

Habitat: This near-shore species occurs on soft bottoms. The stem -like base of the polyp is anchored in the sand or mud ground. This shallow water species occurs up to a depth of 30m.

Biology: It has a long stalk and almost circular frond with deep incision. Its autozooids are few and are well spaced, with three well-developed marginal teeth. Siphonozooids are in groups of 3-7, and these groups are distinctly arranged in rows. Spicules are of the same size; shorter rods (0.25 mm. long) are seen in the stalk; and longer (0.40-0.45 mm.) three- flanged needles are seen in the frond. Color of spicules is purplish and bright yellow. Underside of frond and stalk are pale purplish except for the tip of the stalk. Autozooids and siphonozooids are colorless and are without spicules.All three *viz.* dopamine, noradrenaline, and adrenaline, have been detected to varying degrees in the colonial compartments.

Bioluminescence: In this species, the first sign of bioluminescence followed its settlement, and was delayed if settlement was prevented. Further, bioluminescent responses did not display normal facilitation until the appearance of the primary polyp stage (Satterlie and Case, 1979)

Renilla muelleri

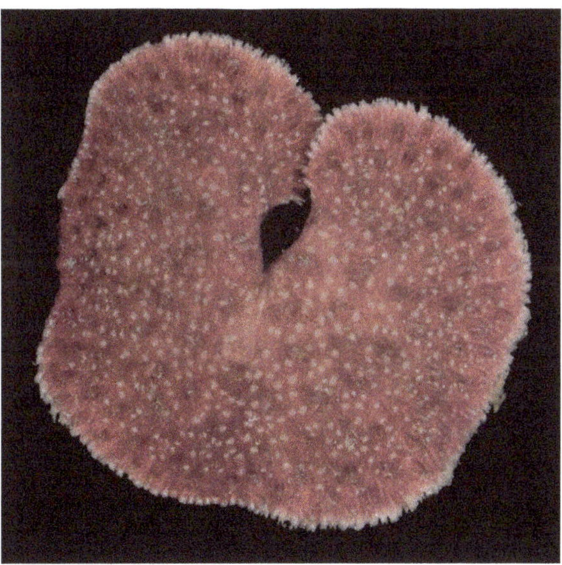

Image credit: Flickr.com

Common name (s): Sea pansy

Global distribution: Gulf Coast of the United States and the eastern coast of South America

Habitat: This euryhaline littoral species is found up to a depth of about 150 m.

Biology: Animals of this species are horseshoe or kidney shaped. Peduncle is relatively short and peduncle sclerites are of the same length as those of the rhachis. Autozooids have five calicinal teeth. In the dorsal section, the rhachis is fully covered with zooids. Ventral part of the rhachis and peduncle are of deep violet coloration.

Bioluminescence: In this species, bioluminescence occurs in a bright wave across the surface, and is generally coordinated by nervous transmission. The bioluminescent wave of this species can be initiated by mechanical, electrical or chemical stimulation. It is also reported that the specialized bioluminescent cells of this species *viz.* photocytes, contain membrane-bound vesicles called luminelles; and when Ca++ ions move into these photocytes, a chemical reaction occurs in which a luciferin substrate reacts with oxygen in the presence the enzyme luciferase to produce light (Goodwin,2006)

Renilla reniformis

Image credit: iNaturalist. (CC)

Common name (s): Sea pansy

Global distribution: It is native to the warm continental shelf waters of the Western Hemisphere. It is also frequently found washed ashore on North East Florida beaches.

Habitat: It has been reported to inhabit a wide range of environments; It may also be found living intertidally completely buried in the sand.

Biology: Animal of this species is round, kidney or heart shaped. It has fairly a long peduncle with very few sclerites. Peduncle sclerites are short and wide and rhachis sclerites are long and thin. Autozooids possess seven calicinal teeth. Medium dorsal tract is free of polyps. Ventral section of rhachis and peduncle are white or pale pink. Sclerites are of various colorations i.e intense violet, yellow, pale pink or white.

Bioluminescence: It is bioluminescent when disturbed and this phenomenon is due to the interplay between a luciferase (Renilla-luciferin 2-monooxygenase) and green fluorescent protein (GFP). Both these molecules have become extremely important in biology (https://en.wikipedia.org/wiki/Sea_pansy).

Biochemically, the bioluminescence of this species has been best studied. This bioluminescence involves two major components *viz.* the luciferin, coelenterazine, and the enzyme Renilla luciferase. These two components are necessary and are believed to be sufficient for light production, though in the natural system, two additional proteins are found. Such proteins are calcium-activated luciferin-binding protein (LBP) for coelenterazine and a green fluorescent protein (GFP) that interacts closely with the luciferase to modify the color of the emitted light by resonant energy transfer (Francis and de Vilar, 2020)

Cormier (1962) reported that the bioluminescent system of this sea pansy involved adenine-containing nucleotides (either adenosine mono-, di-, or triphosphate), an oxidizable substrate (*Renilla* luciferin), molecular oxygen, and an enzyme (*Renilla* luciferase) were required for luminescence in crude extracts of this organism.

Johnsen *et al.* (2012) reported the wavelength of emission maximum of this species as 505 nm

Family: Scleroptilidae

Calibelemnon sp.

Bioluminescence: The emission peak wavelength of this unidentified species was 505 nm (Johnsen *et al.*, 2012)

Family: Stachyptilidae

Stachyptilum superbum

Common name (s): Exquisite sea pen

Global distribution: Northeast Pacific Ocean from central California to the Gulf of Panama; Western Atlantic: USA and Canada.

Habitat: This sessile species is found on the continental shelf; at a depth range of 0-1080 m

Biology: It is a slow growing, long living and large species which can reach up to 25 cm total colony length

Bioluminescence: Johnsen *et al.* (2012) and Widder *et al.* (1983) reported on the emission peak wavelengths of this luminescent species as 535nm and 503-533 nm respectively

Family: Umbellulidae

Umbellula magniflora

Image credit: National Science Foundation. (Applied for permission)

Common name (s): Umbrella sea pen

Global distribution: North-east Atlantic Ocean

Habitat: This sessile species occurs at depths of 4,510–4,860 m

Biology: In this species, its polyps form a rosette at the end of the stalk with an enlargement at its upper end. The terminal polyp is in the centre of a rosette of polypsis disposed in a flattened plane on the dorsal side of the colony. Further, all the secondary polyps are aligned in a single plane, and all polyps are clustered in a single congested polypary.

Bioluminescence: In this luminescent species light emitted from its stalk. Johnsen *et al.* (2012) and Widder *et al.* (1983) reported on the emission peak wavelengths of this luminescent species as 505nm and 470 - 501 nm respectively

Umbellula sp.

Global distribution: The genus *Umbellula* is widely distributed worldwide in all oceans including the Ross Sea

Habitat: Species of *Umbellula* are considered characteristic organisms of deep-sea and abyssal environments. Species of this genus are perhaps the deepest known octocorals and are wide-ranging from about 200 m down to a new record of 6260 m,

Biology: In the species of *Umbellula,* a cluster or whorl of large polyps tops a smooth, long, thin stalk that is devoid of polyps. Colour of the animals is greyish-white when alcohol-preserved. Size of the individuals is highly variable and they are approximately 150–3000 mm in maximum length.

Bioluminescence: The species of *Umbellula* were reported to produce both blue and green light in a tissue-specific manner (Francis and Vilar, 2020).

Bessho-Uehara *et al.* (2020) reported that this unidentified species of *Umbellula* displayed green fluorescence on the tentacles of polyp and basal polyps on the stalk. Further, upon mechanical stimulation, the polyps emitted green bioluminescent light. The peak wavelength of its bioluminescent spectrum was 502 nm.

Image credit: Wikipedia

Cavernularia habereri

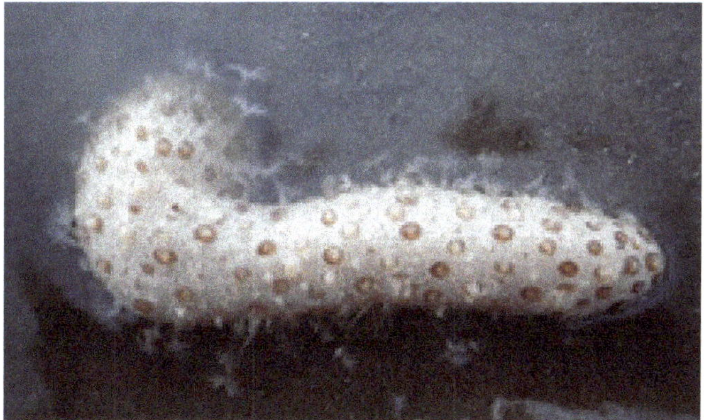

Image credit: iNaturalist. (Applied for permission)

Common name (s): Not reported

Global distribution: It is known to be abundant in the fjord of Aburatsubo, Japan

Habitat: The colony of this species lies hidden in the sand. It gets contracted during the day and it takes up water a night and expands.

Biology: Larger animals of this species have a total length of about 60 cm.

Bioluminescence: Upon stimulation by touching or electrically or by the addition of ammonia, the whole outer surface of this species has been reported to produce a brightly luminous slime. Further, its stalk containing no polyps is especially brilliant (Harvey, 1917).

Cavernularia obesa

Image credit: Wikimedia Commons

Common name (s): Sea cactus

Global distribution: It is widespread throughout the tropical waters of the Indo-Pacific area; Western Pacific: China, Hong Kong, and New Caledonia; India

Habitat: This sessile species is found in shallow waters at a depth range of 5 - 20 m

Biology: Colony of this species is clavate with a total length of 33 - 58 mm. Peduncle is 14 mm long and 10 mm width and is forming about 40 per cent of colony length. Rachis is measuring 19 mm long and 17 mm width. Both rachis and peduncle are thick and fleshy. Autozooids and siphonozooids are distributed on rachis surface with no arrangement. Autozooids are 236 in number and are measuring 2 –3 mm long, 3 mm wide and capable of total retraction into rachis. Tentacles of the autozooids comprise 7–8 arms. Distribution of the autozooids along the apical portion of the rachis are dense and few in number on the lower basal portion. Siphonozooids are densely distributed on rachis surface between the autozooids, and are also covering the distal terminus of the colony

Bioluminescence: The bioluminescence of his species is involves both luciferyl sulfate and photoprotein, and the both these systems afford the light-emitter. The structure of the light-emitter extracted from this species was determined as 2-(p-hydroxyphenylacetamido)-3-benzyl-5-(p-hydroxyphenyl)pyrazine. (Osamu *et al.*, 1975).

Veretillum cynomorium

Common name (s): Round sea pen

Global distribution: It has cosmopolitan distribution

Habitat: This sessile species is collected from circalittoral and bathyal zones

Biology: Species of this genus are cylindrical or slightly clavate. Axis is either well-developed or absent. Polyps are radially arranged throughout the rachis and they are emanating directly from surface of rachis. Polyp leaves are absent. Siphonozooids are numerous between autozooids, and they are sometimes in longitudinal rows. Sclerites are small and are irregularly-shaped plates with sparse denticulation or tuberculation. Sclerites may be absent from the rachis altogether. The specimens of *Veretillum* ranged from 8-12cm when fully contracted and are up to 42cm when they are completely expanded.

Image credit: Wikimedia Commons

Bioluminescence: Specimens of this species had the brightest fluorescence but the dimmest bioluminescence. The light from the species of *Veretillum* was green, though it appeared as blue. Surprisingly, the fluorescence spectrum indicated the most red-shifted fluorescence. In this species, green fluorescence was distinct along the siphonozooids throughout the length of the rachis, as well as the tentacles of the autozooids. Sites of fluorescence are also the same as those emitting light. Light emission however, occurred only in dark-adapted specimens, but animals appeared to be luminous or they could be stimulated regardless of time of day. (Francis and Vilar, 2020)

Henry and Ninio (1978) reported that calcium ions which serve as bioluminescent vesicles can trigger an emission of light from this species. The Ca2+-triggered bioluminescence is stimulated by an asymmetrical distribution of cations or anions; and for the maximal effect either high internal sodium or high external chloride is required.

Family: Virgulariideae

Acanthoptilum album

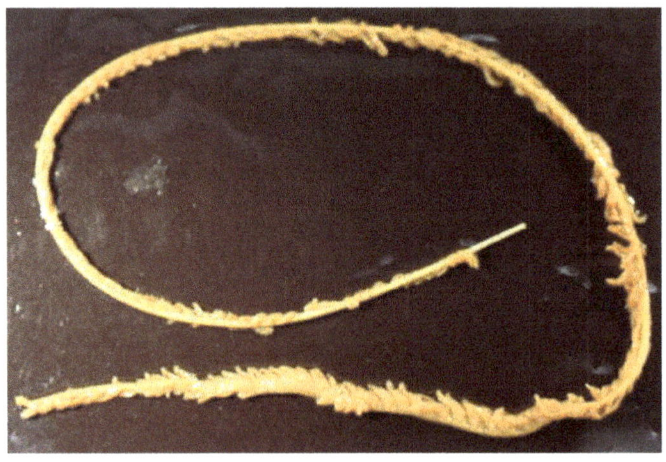

***Acanthoptilum* sp.**

Image credit: Centre for Biodiversity Genomics (No copyright restrictions)

Common name (s): Sea pen

Global distribution: It is widely distributed in the Eastern Pacific.

Habitat: This benthic species has a depth range of 80-142m.

Biology: metric. Colony of this species is very slender, andis bilaterally symmetrical. It is known to attain a height of 225 mm. length of stem to rudimentary leaves is 87 mm. Terminal bulb is not well marked but it is a slight swelling at a distance of 2.5 cm. Polyps are apparently nonretractile and the lobular fringed tentacles are quite well expanded especially in alcoholic conditions. Calyces are much reduced in size and are quite soft,. Zooids are apparently wanting on both dorsal and ventral surfaces of the rachis.

Bioluminescence: Widder *et al.* (1983) reported on the wavelength of emission peak of this luminescent species as 510 nm.

Acanthoptilum annulatum

Common name (s): Not known

Global distribution: U.S. West Coast; California: Monterey Bay

Habitat: This benthic and intertidal species has been reported to have a depth range of 64- 160m.

Biology: Colony of this species is very slender with a length of 156 mm. Stem alone measures 68 mm. There are about 170 pairs of pinna. Full grown pinna measures 5 mm long by 1.5 mm. Polyps are usually with 6 calyces to each well-developed pinna. Longest (distal) polyps are about 2 mm long. Zooids are in groups of 3 to 8 and are laterally placed. Coloration of the colony is pink which is due to the combination of carmine spinules and white coenosarc. Middle part of the stem is purplish, basal part is g light pink, and bulb is whitish.

Bioluminescence: Widder *et al.* (1983) reported on the wavelength of emission peak of this luminescent species as 510 nm.

Stylatula elongata

Common name (s): Sea Whip, Slender sea pen, white sea pen

Global distribution: NE Pacific; Central to Southern California.

Habitat: This sessile is found on sandy and muddy bottoms at depths of 0-60 m. This species stands upright in the mud with its polyps expanded when covered at high tide. At low tide the polyps retreat down the central stalk leaving the bare stalk protruding.

Biology: Colony of this species is scabrous and is more slender form with proportionally large polypiferous lobes and cylindrical stem which is without grooves. Small portion of the stem is bearing the lobes In this species, polyp calyces are separate to the base and there are < 24 polyps per polyp leaf.

Bioluminescence: Upon stimulation, this species glows in dark. Bioluminescence has been reported to occur when a chemical reaction takes place within the sea pen and energy is released in the form of light. Its showy displays resemble the lights on a movie marquee as the flashes circulate up and down the rod-like stem of the sea pen (https://ecology.wa.gov/Blog/Posts/March-2016/Eyes-Under-Puget-Sound-Critter-of-the-Month-%E2%80%94-Slen)

Image credit: Claire Fackler, CINMS, NOAA, Wikimedia Commons

Stylatula sp.

Bioluminescence: Johnsen *et al.* (2012) and Widder *et al.* (1983) reported on the emission peak wavelengths of this unidentified luminescent species as 505nm and 511 nm respectively

Virgularia mirabilis

Common name (s): Slender sea pen

Global distribution: It is found distributed throughout the Mediterranean and Western Europe; also in islands of the Mid-Atlantic.

Habitat: This species is able to live in shallow and sheltered locations, up to depths of 400 m. It stands upright in fine sediments. Due to its location in fine sediments, it is known to be able to withdraw into the sediment very quickly.

Biology: Species of Virgularia are long and whip-like. In this slender, feather-like sea pen, its peduncle gets buried underneath the sediment, while its rachis extends into the water column. It forms elongated and slender colonies, with polyps living on attached side branches. Polyps are able to form groups as large as 16 on each branch. Color of the animal ranges from white to a light yellow. It grows to a maximum length of 60 cm.

Bioluminescence: Nicol 1958) reported that the luminescent response of this species is evoked by mechanical stimulation and it lasts for about one and half seconds. Francis and Vilar (2020) reported that this species emits a blue light.

***Virgularia* sp.**

Image credit: Seascapeza, Wikimedia Commons

Order: Zoantharia Family: Parazoanthidae

Savalia lucifica (= *Parazoanthus lucificum*)

Global distribution: NE Pacific Ocean; off the coast of California, Mediterranean Sea

Habitat: This sessile species is epifaunal on the golden gorgonian, Muricea californica at depths of 0-24 m.

Image credit: Wikiwand

Biology: Colony of this luminescent or yellow zoanthid is large and is consisting of numerous moderately crowded polyps. In life, its largest polyps are 15 mm. long and 3.5 mm. in dia. The holotype colony almost completely covers the numerous branches of a 20 cm. gorgonian

Bioluminescence: This brilliantly bioluminescent species has been reported to secrete copious quantities of thick sticky slime which upon stimulation or when removed from the water is known to emit light. It is also known to emit light when stimulated or when stroked gently by a finger.(https://en.wikipedia.org/wiki/Savalia_lucifica). Its tentacles have also been reported to glow when disturbed.(https://www.marinelifephotography.com/marine/cnidaria/savalia-lucifica.htm). Widder (1983) reported on its wavelengths of emission maximum as 500,502,574nm.

3.3. Nemertean Worm

Ribbon worms (Phylum : Nemertea) are mostly marine, benthic and carnivorous invertebrates. They feed mainly on small polychaete worms, crustaceans and mollusks. Among the nemertean worms, *Emplectonema kandai* is the only currently known bioluminescent member of the genus Emplectonema, and furthermore, it is currently the only known bioluminescent ribbon worm.

Phylum; Nemertea; Class: Hoplonemertea.; Order; Monostilifera;Family: Emplectonematidae

Emplectonema kandai

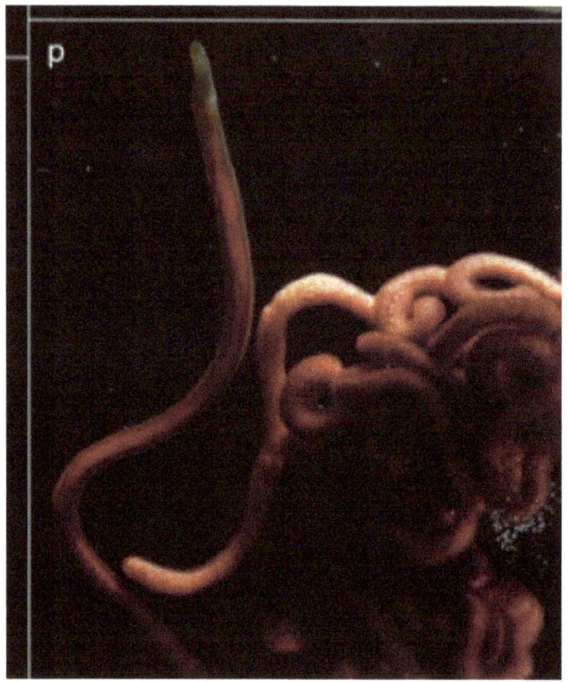

Image credit: Kanda, Sakyo (Applied for permission)

Common name (s): Ribbon worm

Global distribution: The Atlantic and Pacific Oceans, the Mediterranean Sea, the White Sea and Japan.

Habitat: It is a benthic, commensal species and is found by coiling up on the tunicate (sea squirts) Chelyosoma; and it occurs at depths of 35–40 m

Biology: These ribbon worms are reddish orange in color. They possess many eyes and vary in length, from 53–115 cm, and are about 0.5-0.7 mm in dia. These animals depend on ciliary gliding to move around. They are also known to coil up on the wall of a large vat of running sea water or on the bottom, attached by the slime which is abundantly secreted from the surface of its body. If they are not disturbed, they remain in their substrata quietly for two months or more.

Bioluminescence: In this bioluminescent, ribbon worm, the photogenic cells are found distributed throughout the body and these cells appear to have openings in the cuticle. These animals flash brilliantly only on stimulation. Whitish-green light has been reported to appear from the whole of the body, except a small region at the head end, in response to mechanical, chemical, thermal or osmotic stimulation. This light disappears in one or two seconds. In this species luciferin and luciferase could not be demonstrated and there is no extracellular luminous secretion. The origin of the light is therefore thought to be the gland cells of the worm and not symbiotic bacteria(Kanda,1939; http://ecoursesonline.iasri.res.in/mod/page/view.php?id=86809; http://photobiology.info/Haddock.html)

3.4. Peanut Worm (Sipunculan worm)

The "peanut worms" take the name due to the general shape of shelled peanuts. There are about 320 species of sipunculan worms which are all marine and mostly from shallow waters. While some (like Sipunculus) burrow into sand and mud, others live in crevices in rocks, or in empty shells. Still others bore into rock. Among these species, only one species *viz. Nephasoma pellucidum* has been reported to be bioluminescent

Phylum: Sipuncula; Class: Sipunculidea; Order: Golfingiiformes; Family: Golfingiidae

Nephasoma pellucidum

Image credit: National Science Foundation (CC)

Common name (s): Sipunculan worm

Global distribution: Indo-Pacific and Western Atlantic; from Australia and Indonesia to southern Japan.

Habitat: This benthic and subtidal species inhabits sandy bottoms and dead corals at depths of 4 - 200 m. It is a facultative mobile burrower.

Biology: Body shape of this species is vermiform. Trunk is spindle-shaped. Introvert is shorter than trunk, and is with scattered hooks of two types – small, about 50 µm, and large, up to 300 µm in height. Tentacular crown has 20–30 whitish transparent tentacles. Trunk is translucent,and pale tan and is with large uniformly distributed digitiform papillae. This species grows to a maximum length of 25 mm and width of 4 mm.

Bioluminescence: It is a new record of bioluminescence. Individuals of this species emitted green luminescence. Each individual produced a pulse of visible greenish light upon perturbations with a blunt probe. Further, it was found that the glow was localized on its body wall with no apparent patterning. The intensity was not throughout the body. Its proteins had similarities to *Renilla*-type luciferases (Oliviera *et al.*, 2021)

3.5. Polychaetous Annelids

In the phylum Annelida, about 100 species are known to be bioluminescent and all such species are reported to date are marine polychaetes. Bioluminescent polychaetes are found in a wide range of marine habitats from intertidal and shallow waters to the deep-sea; both in the benthic and pelagic zones,; and from polar to tropical regions. These animals are characterized by having parapodia with chitinous chaetae. Bioluminescence in polychaetes is present in several species distributed in eight families including Acrocirridae, Alciopidae, Chaetopteridae, Cirratulidae, Flabelligeridae, Polynoidae, Syllidae, Tomopteridae, and Terebellidae. Annelids are known to luminesce in three colors: yellow, blue, and green. Most polynoids display a green luminescence, which is rarely found in the deep sea. Yellow luminescence is rare and is found in the planktonic *Tomopteris* species while green light is the second most common color in annelids (Moraes *et al.*, 2021).Light production in these polychaetes is very diverse, with emission wavelengths ranging from 445 nm in Terebellidae to 573 nm in Tomopteridae. While *Chaetopterus* and swarming females of *Odontosyllis* display long glows emitted from secretions, polynoids and swarming males of *Odontosyllis* emit flashes of light via intracellular bioluminescence. (Verdes and Gruber, 2017)

Phylum: Annelida; Class: Polychaeta; Order: Canalipalpata; Family: Chaetopteridae

The Chaetopteridae which is a polychaete family include tubicolous animals,commonly occurring on muddy or sandy mud bottoms from intertidal to shallow shelf waters of temperate and tropical seas. In this family, there are

65 currently accepted species which are placed in 4 genera *viz. Chaetopterus, Mesochaetopterus, Phyllochaetopterus,* and *Spiochaetopterus*. These chaetopterids are well known as hosts harbouring numerous symbiotic associates (including complex communities) inside their parchment-like tubes. These tubes provide shelter with continuous water flow which brings oxygen and food items to such symbionts. Among the chaetopterids, the genera, *Chaetopterus,* and *Mesochaetopterus* possess luminescent representatives

Chaetopterus cautus

Image credit: Alexander Semenov, Reproduced with permission

Common name (s): Not reported

Global distribution: Japan

Habitat: This species has been reported to harbour pea crab Pinnixa banzu in its tube in the intertidal sandy bottom

Biology: Not reported

Bioluminescence: Kin and Oba (2020) reported that this species emitted blue light under dark field after the stimulation with potassium chloride solution (400 mM) and freshwater. Its emission peak was at the wavelength of 459nm. It is also reported that its souped-up ferritin protein manages iron metabolism in cells by storing and releasing it in a controlled way.

Amazingly, the species of *Chaetopterus* can keep their bioluminescent blue light glowing for hours and sometimes even for days. Further, the worms' bioluminescence depends largely on the iron dissolved in seawater; and the worm's mucus contains iron which regulates its light production. However,

Chaetopterus kept in seawater without iron can still secrete mucus with strong luminescence (https://today.ucsd.edu/story/bioluminescent_worm_found_to_have_iron_superpowers)

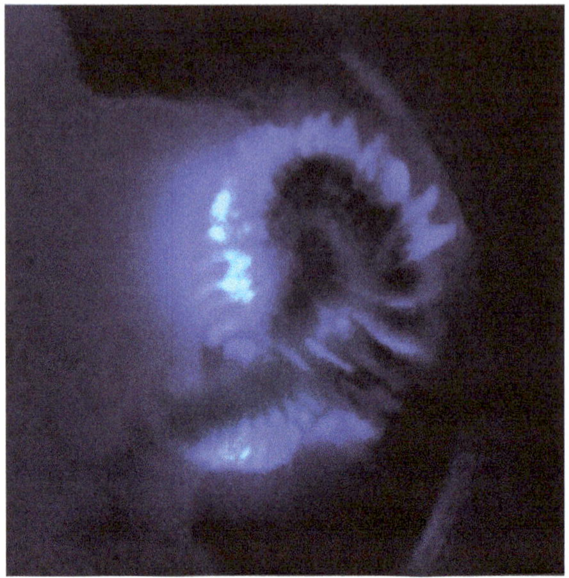

Image credit: Yuichi Oba, Reproduced with permission

Chaetopterus dewysee

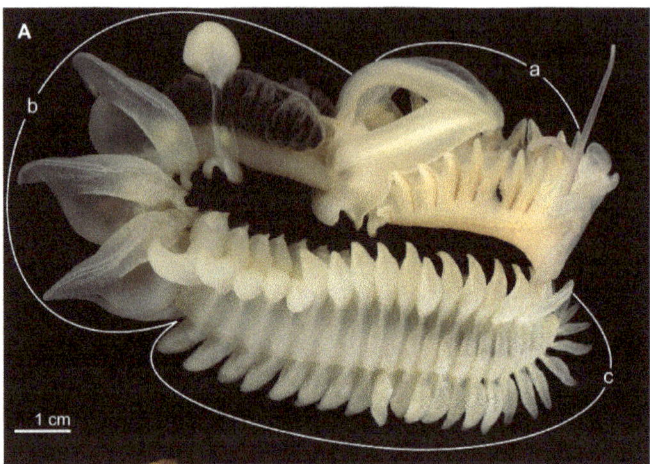

Image credit: Gregory Rouse, Reproduced with permission

Common name (s): It is a new species

Global distribution: Southern California.

Habitat: It is commonly found partially buried along canyon walls in large groups of solitary, intermingled tubes and sediment with other fauna like sponges and tunicates. Its parchment like and U-shaped tube sand debris and shell fragments on the outer surface. Total tube length is 293 mm

Biology: It has a total body length of 55 mm. Region A is with 10 chaetigers. Prostomium is small, with an entirely rounded anterior border. Peristomium is extended and is completely covering prostomium. Two grooved palps are found extending beyond peristomium. Its paired eyes are located at the base of palps.

Bioluminescence: This species has served as a model for investigating the biochemistry of light production and the bioluminescent properties of *Chaetopterus mucus* (Tilic and Rouse, 2020).

Chaetopterus gregarius

Common name (s): Not known as it is a new species

Global distribution: Tokyo Bay, Central Japan

Habitat: This benthic, tubicolous organism is is gregarious and probably epifaunal and is commonly found from hard substrata in shallow waters at depths of 120-300 m.

Biology: Its tough and parchment-like tube is U-shaped. Body of the species is oculate and of moderate size (up to 57mm long), with 9A+5B+16-23C segments. It differs from other members of the genus in having two types of notopodia in region C.

Bioluminescence: It has been reported to secrete a steadily luminous slime from various parts of its body and emit a flash of light from the posterior segments. This light may last a few seconds (Verdes and Gruber, 2017). It is also reported that when live animals were placed in fresh water, its bioluminescence was seen in the palps and alate (winglike) notopodia of B1. Most long-lasting and brilliant light however appeared at the base of the B1 notopodia and no other areas luminesced.(https://irdb.nii.ac.jp/en/00822/0001821732)

Chaetopterus variopedatus.

Common name (s): Not reported

Global distribution: This cosmopolitan species is found in temperate and tropical locations throughout the worldwide. It is very common in the eastern Pacific.

Habitat: This benthic, tubicolous organisms is commonly found in shallow coastal habitats. It is known to build and live permanently in a tough, flexible, papery U-shaped tube which is buried in soft substrate with both ends protruding like little chimneys.

Image credit: Smithsonian Tropical Research Institute (Applied for permission)

Biology: This segmented, pale coloured worm grows up to a length of 25 cm. Anterior end of the worm is short and it has bristle-bearing segments and a shovel-like mouth. Middle section bears parapodia. On the 12th segment these parapodia are modified into long wing-like structures which are known to secrete mucus and form a bag. Parapodia on segments 13, 14 and 15 get fused into three paddle-shaped, piston-like structures, Posterior half of the worm is segmented and tapers towards the rear.

Bioluminescence: It has been reported to emit transitory undispersed light and secrete luminescent mucus that is scattered in the surrounding water when direct mechanical stimulation or freshwater is applied to the peristomial palps, dorsal surface or notopodia. The components of its bioluminescence system were Riboflavin/luciferase/H2O2/Fe2+ and its emission maximum wavelength was 463 nm(Verdes and Gruber, 2017). It is also reported that its intrinsic bioluminescent emissions can be observed from its tube as a blue light which is emitted from the extremities of individual body segments, or as a secreted mucus. (https://www.google.com/search?q=bioluminescence+in+Chaetopterus and oq=bioluminescence+in+Chaetopterus and aqs=chrome.69i57j33i160l3.5469j 1j15 and sourceid=chrome and ie=UTF-8).

Mesochaetopterus japonicus

Common name (s): Not reported

Global distribution: West Pacific; Japan,China and Taiwan

Habitat: This benthic, tubicolous organism is commonly found in shallow,intertidal habitats, protected beaches, and estuaries. It harbours the symbiotic pinnotherid crab *Tritodynamia rathuni* inside its l-shaped tube.

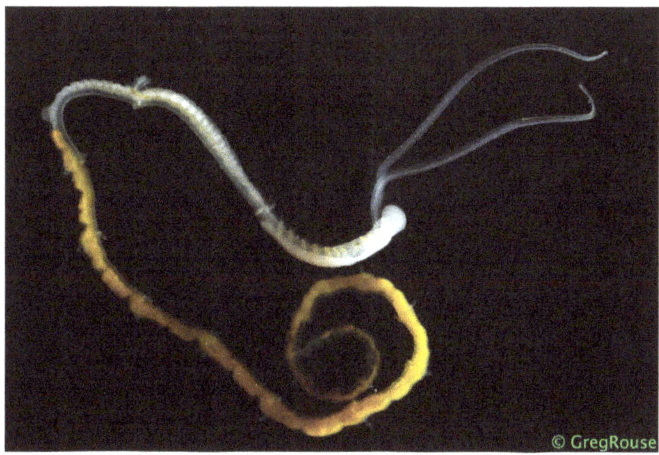

***Mesochaetopterus* sp.**

Image credit: Greg Rouse. (Applied for permission)

Biology: Mesochaetopterus can be distinguished from the other genera of chaetopterids by the presence of large palps and non-fused notopodia, and absence of tentacular cirri. Anterior region of the trunk of this species is cream-colored. It has large wing-shaped notopodia and palps do not bear longitudinal colour stripes.

Bioluminescence: Kin and Oba (2020) reported that this species emitted blue light under dark field after the stimulation with potassium chloride solution (400 mM) and freshwater. Its emission peak was at the wavelength of 460 nm

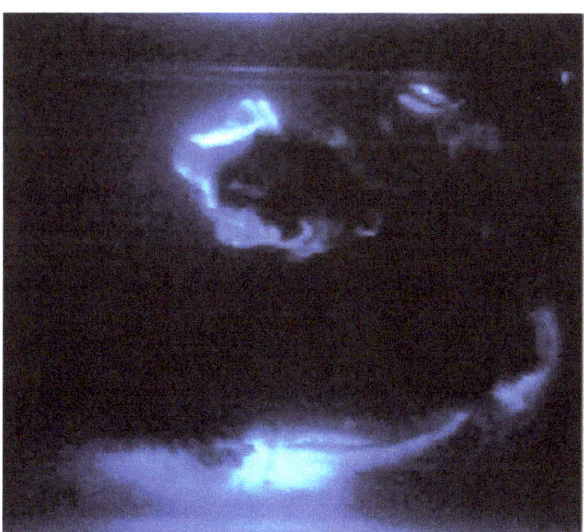

Bioluminescence under Dark Field after KCl Stimulation

Image credit: Yuichi Oba, Reproduced with permission

Order: Eunicida; Family: Onuphidae

Onuphis sp.

Image credit: Ecology WA, Wikimedia Commons

Common name (s): Not known

Global distribution: Eastern Atlantic

Habitat: This worm lives in tubes which are constructed of parchment-like material secreted by it.

Biology: Body of this species is elongated and is with numerous segments and with an iridescent cuticle. Prostomium is oval and is with two globular ventral palps and two oval frontal palps. Its strong, muscular proboscis has a jaw apparatus consisting of two mandibles and seven or nine maxillae.

Bioluminescence: It has several pairs of lateral luminous spots in each segment that emit a bluish-green light when stimulated (Verdes and Gruber, 2017)

Order: Phyllodocida; Family: Alciopidae

Alciopina sp.

Common name (s): Not reported

Global distribution: Mexico

Habitat: Not reported

Biology: Body of this species is short and its prostomium is rounded. It has one ventral pair of palps and three antennae (one dorsal pair plus a median one which is represented by a small process). Pharynx is short, with marginal papillae. Eyes are large, and are laterally directed. There are six pairs of tentacular cirri. Chaetae are simple, capillary, and acicular on anterior segments. Segmental glands are seen from first chaetigerous segment.

Bioluminescence: It is a luminescent species (http://species-identification.org/species.php?species_group=zsao and menuentry=inleiding and record=Po.%201%20). No other information.

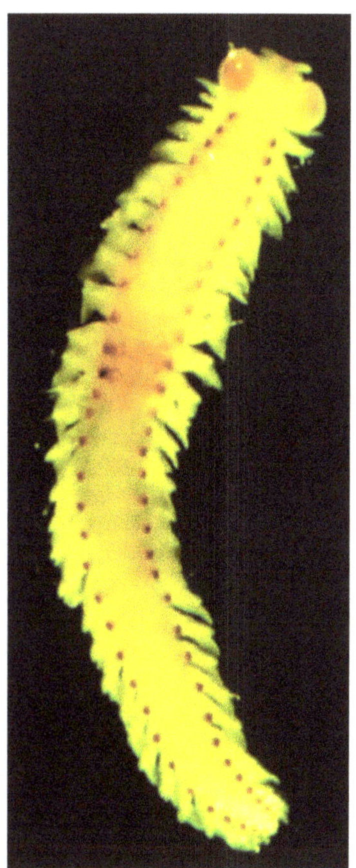

Image credit: María Ana Fernández-Álamo (Applied for permission)

Krohnia lepidota (= *Alciope lepidota*)

Common name (s): Not known

Global distribution: Western Central Atlantic and Pacific Ocean.

Habitat: This benthic burrowing species is seen at depths of 90-150m.

Biology: It has transparent body and its prostomium has a pair of very large globular eyes, and 4-6 small antennae. Parapodia are uniramous.

Bioluminescence: León-González *et al.* (2021) reported that its bioluminescence is due to its pigmented segmental glands serving as photogenic organs.

Rhynchonereella angelini

Image credit: Fisheries and Oceans Canada, Moira Galbraith, WoRMS (Reproduced with permission)

Common name (s): Not known

Global distribution: Western Central Atlantic and Eastern Pacific.

Habitat: This a benthic and holopelagic species.

Biology: Its pharynx is muscular, long, and is about eight times longer than the eye diameter. There are five tentacular cirri. Parapodia have foliaceous dorsal and ventral cirri. Anterior parapodia possess 14–18 acicular chaetae, plus some capillary chaetae.

Bioluminescence: This species has been reported to possess mucus glands producing luminescent material (http://species-identification.org/species.php?species_group=zsao and menuentry=inleiding and record=Po.%201%20

Order: Phyllodocida; Family: Polynoidae

Acholoe squamosa (= *Acholoe astericola*)

Common name (s): Not known

Global distribution: It occurs in the Mediterranean Sea and North-east Atlantic Ocean

Habitat: It is a commensal species with the burrowing Starfish, *Astropecten irregularis*. Most adult worms occurred in the ambulacral grooves of this starfish.

Biology: Body of this species is long, with numerous segments (50 -100) and numerous pairs of elytra on segments 2, 4, 5, then on alternate segments to 23 and on every third segment thereafter. Prostomium is anteriorly rounded (without peaks) and the lateral antennae are inserted terminoventrally. Parapodia have elongate acicular lobes with both acicula penetrating epidermis. Notochaetae are stout, and are with distinct rows of spines and blunt tips. It grows to >15 mm in length.

Image credit: Asternatura. (Applied for permission)

Bioluminescence: The bioluminescent scales of this species are covered with photogenic epithelial cells (photocytes) in which the increase of the stimulus strength elicits another action potential specifically correlated with a flash. The luminescent activity is intracellular and occurs in brief flashes (Bilbaut,1980).

Verdes and Gruber (2017) reported that the ventral epithelium of the elytra of this species has a layer of luminescent cells or photocytes, which are modified epidermal cells

Nicol (1953) reported that the photogenic cells of this species are tall, polygonal in cross section, and are containing a large oval nucleus. Proximally, these cells are filled with large oval or round granules which are considered to be protein in nature, and are representing the luminescent material. Further, he reported that when this animal is quickly cut into two, only the elytra of the posterior portion emits light and the anterior half remains dark.

Moraes *et al.* (2021) reported that the wavelength of emission maximum of this species was 520 nm.

Eunoe nodosa

Common name (s): Not known

Global distribution: Arctic, North Pacific, North Atlantic to English Channel

Habitat: This sublittoral species occurs up to a depth of 1260 m, amongst algae or on hard bottoms. It is also a commensal species with pagurid crustaceans

Biology: It has an elliptical body in which the dorsal and ventral surfaces are smooth.

***Eunoe* sp.**

Image credit: Polychaete Party, Wikimedia Commons

Prostomium is bilobed, and is with a peak on each lobe, a median antenna, a pair of lateral antennae, and a pair of palps with 6 rows of papillae. There are two pairs of eyes. Scales which are thick and kidney-shaped overlap and are completely covering the body. Notopodia are small mounds with a ventral acicular ligule, and numerous chaetae are seen in regular rows. Notopodial chaetae are stout and neuropodia are well developed. Animals with 37 chaetigers grow up to 90 mm.

Bioluminescence: It is a bioluminescent species (Nicol,1953). No other information

Gattyana cirrhosa

Image credit: Picton B.E. and Morrow,C.C. (Applied for permission)

Common name (s): Not known

Global distribution: Arctic, Northern Atlantic, Mediterranean Sea and Northern Pacific.

Habitat: It occurs in muddy substrates from the intertidal zone to depths of about 1,600 m. It has a commensal relationship with chaetopterid, terebellid, and pectinariid polychaete worms, which are living within the tube it constructs. However, it is also a free-living taxon and it is often found in polychaete tubes, *e.g. Amphitrite, Chaetopterus, Thelepus* or with bryozoans *Melobesia* or *Amphihelia.*

Biology: It is a short-bodied, pale yellow to brown worm with 38 segments and 15 pairs of elytra, which bear a marginal fringe of papillae. Its dorsal and ventral surfaces are smooth. Lateral antennae are positioned ventrally on the prostomium, directly beneath the median antenna. There are two pairs of eyes; anterior pair is located below the prostomial peaks; and the posterior pair is in front of the rear margin. Animals with 38 chaetigers grow up to 50 mm.

Bioluminescence: Moraes, *et al.* (2021) and Verdes and Gruber (2017) reported that the wavelength of the emission maximum in this species was 515 nm and 517 nm respectively

Harmothoe aculeata

Image credit: Bidiversidadmexicana. (Applied for permission)

Common name (s): Not reported

Global distribution: Western Central Atlantic: Central America to northern Brazil.

Habitat: This benthic and intertidal species occurs up to depth of 69 m

Biology: In this species, total number of setigerous segments varies from 34 to 36. Spines of the first pair of scales are shorter than those more posteriorly. A characteristic feature is the neuropodial lobe, which is prolonged into a slender, dorsal, attenuated tip in this species.

Bioluminescence: In all the luminescent species of *Harmothoe*, an area of the epithelium of the lower surface of the elytra which is containing a layer of luminescent cells or photocytes (actually modified epidermal cells), has been reported to emit light flashes upon stimulation. This bioluminescence is actually originated in a membrane photoprotein called polynoidin that reacts specifically to the presence of superoxide anions (a reactive oxygen species, ROS). It is also suggested that this signal could be either a warning or a distracting mechanism. (Pluscheva and Matin, 2009; Moraes *et al.*, 2021)

Anterior Region Displaying Green Bioluminescence by Mechanical Stimulation.

Image credit: Anja Schulze (Applied for permission)

Harmothoe areolata

Image credit: iNaturalist (CC)

Common name (s): Not reported

Global distribution: Subtropical; Northeast Atlantic and Mediterranean Sea

Habitat: This benthic species occurs between stones, algae and the marine plant Posidonia rhizomes at depths up to about 100 m.

Biology: It has an anterior pair of eyes which are anteroventral in position. Dorsal cirri are papillate and are of two alternating kinds: *viz.* inflated subdistally to filiform tip or tapering. Body is with 37 segments. At anterior end its prostomium is bilobed, with distinct cephalic peaks. There are 15 pairs of elytra, covering dorsum, on segments 2, 4, 5, 7, then on every second segment to 23, 26, 29, 32.Parapodia are biramous and neuropodia are with elongate prechaetal acicular lobe with digitiform supra-acicular process. A 37segmented animals has an average length of 18 mm.

Bioluminescence: In this species, the photoprotein polynoidin isolated from its luminescent organ *viz.* elytra emitted light and the wavelengths of maximum emission were 515 nm and 695 nm (Plyuscheva and Martin, 2009).

Harmothoe aspera

Common name (s): Not known

Global distribution: Arctic, North Pacific, North Atlantic to Spain, northern North Sea, Skagerrak and Kattegat.

Habitat : Individuals of this species have been reported to occur in fouling communities growing on steel in vertical orientation

Biology: Prostomium of this scale-worm is slightly longer than broad, with a median antenna, a pair of lateral antennae, and a pair of palps. There are two pairs of eyes. While the anterior pair is on the line of greatest width, the posterior pair is near the rear margin. Scales overlap, and are often leaving a few posterior segments uncovered. First pair of scales is round, and the other pairs are oval to kidney-shaped. At least half of the margins of these scales are fringed with short papillae, Notopodia are small mounds with a ventral acicular ligule and many chaetae. Notopodial chaetae are spinose with blunt tips.

Neuropodia are well developed with an anterior ligule and chaetae. Living animal is greenish and scales are with brown spots.An animal with 34 chaetigers has a length of 30 mm

Bioluminescence: Moraes *et al.* (2021) reported that it is a bioluminescent species. No other information

Harmothoe extenuata (= *Lagisca extenuata*)

Common name (s): Yellow and brown scaleworm

Global distribution: Arctic, Atlantic Ocean, the Mediterranean and Pacific Ocean.

Habitat: This benthic and littoral species is found on rocky shores at depths of 0- 1829m. Sometimes it may be found in the tubes of other polychaetes.

Image credit: Flickr.com. (Applied for permission)

Biology: It is a scale-worm with a body of uniform width. Dorsal and ventral surfaces are smooth. Prostomium is bilobed, with a peak on each lobe, a median antenna, a pair of lateral antennae, and a pair of palps. Of its two pairs of eyes, the anterior pair is on the line of greatest width and the posterior pair is in front of the rear margin. First segment is bearing chaetae and a pair of dorsal and ventral tentacular cirri. Scales overlap, but they leave some posterior segments exposed. First pair of scales are round, and the rest vary from oval to kidney-shaped. Margins of scales with very short papillae on less than a quarter of the circumference. Notopodia are small mounds and notopodial chaetae are spinose with blunt tips. Neuropodia are well developed with an anterior acicular ligule and numerous chaetae. Neuropodial chaetae are spinose with bidentate tips and a few upper and lower chaetae are spinose with unidentate tips. Body coloration is from grey to red with plain or patterned. An animal with 47 chaetigers grows up to 70 mm.

Bioluminescence: Nicol (1953) reported that when this animal is quickly cut into two, only the elytra of the posterior portion emits light and the anterior half remains dark. Moraes etal.(2021) reported its wavelength of emission maximum as 515 nm

Harmothoe imbricata

Common name (s): Not known

Global distribution: It is found throughout the northern hemisphere; and from Alaska to California

Image credit: Eric A. Lazo-Wasem - Gall L, Wikimedia Commons

Habitat: It is a free living species intertidally under rocks and in eelgrass; and subtidally in kelp holdfasts or mussel beds. It may also live commensally with echinoderms or other polychaetes such as *Thelepus crispus, Neoamphitrite robusta,* and *Diopatra ornata*. Its depth range of occurrence is 0–3710 m.

Biology: Like other members of this family, the dorsal side of this species is covered with a series of platelike 15 pairs of elytra which cover nearly all the segments. These elytra have a light fringe of papillae around the edge. There is only one type of notoseta and one type of neuroseta on the segment. Most of the neurosetae fork at the tip and are thicker than the neurosetae. One pair of eyes is visible dorsally. Its lateral prostomial antennae are found inserted ventrally to the medial antenna. Color of the animal is highly variable, though brown is common.

Bioluminescence: As in other polynoids, the light appears in this species in flashes or scintillations when the animal is irritated (Nicol,1953). Plyuscheva and Martin (2009) reported that in this species, the photoprotein polynoidin isolated from its luminescent organ *viz.* elytra emitted light and the wavelength of maximum emission was 520nm. Verdes and Gruber (2017) observed that the luminescence of this species is due to the ventral epithelium of its elytra containing a layer of luminescent cells or photocytes, which are modified epidermal cells

Harmothoe impar

Common name (s): Not known

Global distribution: Eastern Atlantic from the British Isles and Senegal; Mediterranean Sea to the Suez Channel; and the Black Sea.

Habitat: Littoral on rocky shores, sublittoral to 655 m on a variety of bottoms.

Biology: It is a scale-worm with a body of uniform width. Dorsal and ventral surfaces are smooth.Prostomium is bilobed, with a peak on each lobe, a median antenna, a pair of lateral antennae, and a pair of palps. Two pairs of eyes are seen. First segment bears chaetae and a pair of dorsal and ventral tentacular cirri. Scales overlap, covering the body. First pair of scales are round and remaining scales are oval to kidney-shaped. Notopodia are small mounds with a ventral acicular ligule and many chaetae. Notopodial chaetae are spinose with blunt tips. Neuropodia are well developed with an anterior ligule and chaetae. Scales of this species are brownish and sometimes with one or more dark patches. Animal with 40 chaetigers has a maximum length of 25 mm.

Image credit: Fredrik Pleijel (Applied for permission)

Bioluminescence: The bioluminescence of this species is due to its photogenic cells which are tall and are containing a large oval nucleus. Proximally these photogenic cells are filled with large oval or round granules which are closely packed together. These granules are of protein in nature, and are considered as representing the luminescent material (Nicol,1953; Moraes *et al.,* 2021)

Harmothoe longisetis

Common name (s): Not known

Global distribution: Northeast Atlantic and the Mediterranean.

Habitat: This benthic species occurs on hard substrates.

Biology: Its anterior pair of eyes are dorsolateral and are located at the widest part of prostomium. There are 15 pairs of elytra,the margin of which is smooth. At anterior end, prostomium is bilobed. Noto- and neurochaeta have distinct rows of spines. Tip of neurochaetae is straight or only slightly hooked. Neuropodial supra-acicular process are slender and digitiform. Body has 38 segments. At anterior end prostomium is bilobed. Parapodia are biramous. Animal with 38 segments grow up to 30 mm.

Bioluminescence: The wavelength of the emission maximum of this species has been reported to be 515 nm (Moraes *et al.,* 2021) and 512 nm (Verdes and Gruber,2017).

Harmothoe spinifera (= *Polynoe torquata*)

Common name (s): Not known

Global distribution: Northern Spain and the Mediterranean, western English Channel, Irish Sea, western Ireland, Shetland.

Habitat: This littoral and sublittoral species inhabits the holdfasts of seaweed, Laminaria. It is also reported from the burrows of other polychaetes such as *Perinereis cultrifera* and *Nereis irrorata.*

Biology: It is a scale-worm with a body of uniform width. Dorsal and ventral surfaces are smooth. Prostomium is bilobed, with distinct peaks on each lobe, a papillated median antenna, a pair of lateral antennae, and a pair of palps. Of its two pairs of eyes, the anterior pair is seen below the prostomial peaks and the posterior pair is in front of the rear margin. Scales overlap, covering the body. Its first pair of scales are round and the remaining scales are oval to kidney-shaped. Notopodia are small mounds with a ventral acicular ligule and abundant chaetae. Notopodial chaetae are stout and spinose with blunt cleft tips. Neuropodia are well developed with an anterior acicular ligule. Neuropodial chaetae are spinose with bidentate tips. Colour of the scales is brown to black. An animal with 39 chaetigers has a maximum length of 15 mm.

Bioluminescence: Its luminescence is due to its photogenic tissue which is present in the region of the scale around the elytrophore and is consisting of a layer of mucus-cells (Nicol,1953; Moraes *et al.*, 2021)

Lepidasthenia stylolepis

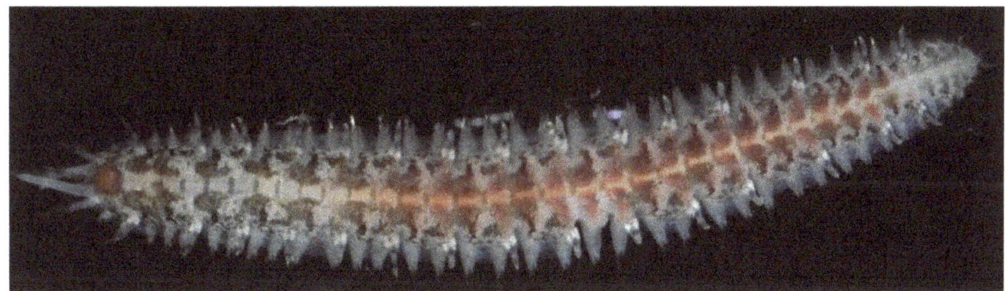

Image credit: EOL (CC)

*Common name (s):*Not recorded

Global distribution: Tropical; Western Indian Ocean: Persian Gulf.

Habitat: This benthic, burrowing species inhabits silty sand. It is also a commensal species associated with the sipunculid *Siphonosoma vastus.*

Biology: This vermiform shaped species grows to a maximum length of 100 mm.

Bioluminescence: It is a bioluminescent species (Moraes *et al.*, 2021)

Lepidonotus clava

Image credit: Jean Michel Crouzet, Reproduced with permission

Common name (s): Button scale worm

Global distribution: Indian Ocean, Red Sea, North Pacific, North Atlantic up to West Africa, Mediterranean, Adriatic, English Channel and southern North Sea

Habitat: This littoral and sublittoral species occurs usually under stones and in algal holdfasts.

Biology: Body of this scale-worm is of uniform width. Dorsal and ventral surfaces are smooth.

Prostomium is bilobed with median and lateral antennae. Of its two pairs of eyes, anterior pair is at the extreme lateral edges of prostomium, and posterior pair is closer together at the posterior margin. First segment is bearing 4 or 5 chaetae and a pair of dorsal and ventral tentacular cirri. Scales overlap in the anterior region. First pair of scales is round, and the remaining ones are oval to round. Notopodia are small mounds with a ventral acicular ligule and numerous chaetae. Notopodial chaetae are short and spinose with blunt tips. Neuropodia are well developed with chaetae. Color of the animal is variable and scales are light brown and are often with darker markings A 26 segmented worm grows to a maximum length of 45 mm.

Bioluminescence: In this species, the photoprotein polynoidin isolated from its luminescent organ *viz.* elytra emitted light and the wavelengths of maximum emission were 525 nm and 620 nm (Plyuscheva and Martin, 2009)

Lepidonotus squamatus (= *Polynoe squamatus*)

Image credit: Eric A. Lazo-Wasem - Gall L, Wikimedia Commons

Common name (s): Scale worm

Global distribution: Atlantic and the Pacific Oceans

Habitat: It occurs in the sublittoral and littoral zones at depths down to about 2,700 m Its habitat is typically beneath stones or among tangled growth.

Biology: Prostomium of this species has two lobes and bears several pairs of antennae, a pair of palps and two pairs of eyes. Dorsal surface of the body, which has uniform width, is completely concealed by two rows of overlapping scales. Its scales are modified cirri and are supported on short stalks. They are covered in tubercles of varying sizes, and have a fringe of papillae. This worm has 26 segments and grows to a length of about 5 cm. It is often found covered with mud

Bioluminescence: The scales of this species have been reported produce a faint bioluminescent glow. (https://en.wikipedia.org/wiki/Lepidonotus_squamatus). Under experimental conditions, homogenized elytra of this species emitted an increasing luminescence in the presence of a superoxide donor system and the wavelength of maximum emission was 520nm (Plyuscheva and Martin, 2009)

Malmgrenia castanea

Common name (s): Chestnut scale worm

Global distribution: Arctic, Northeast Atlantic, Mediterranean, Adriatic, Red Sea, English Channel, North Sea.

Habitat: This species inhabits in benthic areas from sublittoral to 200 m depth, and is usually associated with echinoderms, mostly *Spatangus purpureus* or *Astropecten irregularis.*

Biology: It is a scale-worm with a body of uniform width. Dorsal and ventral surfaces are smooth. Prostomium is bilobed and is without peaks but with a median antenna, a pair of lateral antennae, and a pair of palps. Of its wo pairs of eyes, anterior pair is in front of the line of greatest width, and the posterior pair is well in front of the rear margin. First segment bears chaetae and a pair of dorsal and ventral tentacular cirri. Scales overlap, almost covering the body. First pair is round, and the remaining scales are oval to kidney-shaped.

Image credit: Jean Michel Crouzet, Reproduced with permission

Notopodia are small mounds with a ventral acicular ligule and few chaetae. Notopodial chaetae are spinose with blunt tips. Neuropodia are well developed with an anterior acicular ligule and few chaetae. A worm with 38 segments grows up to a length of 18 mm.

Bioluminescence: It is a bioluminescent species (Moraes *et al.*, 2021). Nicol (1953) reported that when this animal is quickly cut into two, only the elytra of the posterior portion emits light and the anterior half remains dark.

Malmgrenia lunulata (= *Harmothoe lunulata*)

Common name (s): Not reported

Global distribution: Mediterranean

Habitat: It is a crawling species.

Biology: In this vermiform shaped species, its lateral antennae are terminoventral in position. Notochaetae are with entire tip and all neurochaetae are stout. Individuals can grow up to a total length of 100 mm.

Bioluminescence: In this species, light appears in flashes or scintillations when the animal is irritated. (Nicol,1953). Verdes and Gruber (2017) reported that in this species, the ventral epithelium of the elytra has a layer of polynoidin producing luminescent cells or photocytes, which are modified epidermal cells. The components of bioluminescence system in this species are polynoidin/ superoxide anions/O2. Moraes *et al.* (2021) reported that the wavelength of maximum emission in this species was 510 nm.

Image credit: The Marine Life Information Network. (CC)

Neopolynoe acanellae

Common name (s): Scale worm

Global distribution: Northwestern and Northeastern Atlantic Ocean

Habitat: This species inhabits at depths of 400 to 2000 m and is found attached to its host, the pennatulacean coral (sea pen) *Pennatula grandis.* It has also a commensal relationship with host corals of the genus Anthomastus and Acanella, as well as with sponges

Biology: Neopolynoe acanellae has up to 72 segments with 15 pairs of elytra that bear marginal fringe of papillae but no color patterning. Lateral antennae are positioned ventrally on the prostomium, directly beneath the median antenna ceratophore and almost obscured in dorsal view. The notochaetae are about as thick as neurochaetae and only possess simple tips.

Bioluminescence: In this species, its luminescence is due to presence of brighter cells near the elytrophore scar in the elytron. The emission spectrum of this species showed a maximum emission peak at 530 nm- (Taboada *et al.,* 2020).

Neopolynoe chondrocladiae

Common name (s): Scale worm

Global distribution: North-east Atlantic Ocean

Habitat: This luminous deepsea polynoid which lives at depths of about 700 to 2500 m has a unique and obligate symbiotic relationship with the carnivorous sponges *Chondrocladia robertballardi* and *Chondrocladia virgata.*

Biology: This species may have up to 94 segments. Body of this polynoid is oval, and is dorsoventrally flattened with alternating 15 pairs of elytra covering the dorsum These elytra invariably possess papillae, tubercles or ridges adorning the margins and are circular to renal form in shape. Its lateral antennae are found inserted ventrally to prostomium, directly beneath the median antenna. Notochaetae are distinctly thicker than the neurochaetae and possess only simple tips.

Bioluminescence: This species has been reported to be the only bioluminescent deep sea polynoid (Moraes *et al.,* 2021).Taboada *et al.* (2020) reported that the bioluminescence of this species is produced by its elytra which have calix-shaped photocytes arranged along their ventral side. Further, brighter fluorescent cells are arranged near-concentrically around the elytrophore scar and concentrated toward the centre of the elytron. The emission spectrum of this region showed a maximum emission peak at 525 nm, and a smaller peak at 580 nm.

Polynoe scolopendrina

Image credit: iNaturalist. (Applied for permission)

Common name (s): Not known

Global distribution: Atlantic between Iceland and South Africa; Mediterranean, Adriatic, Black Sea, English Channel, northern North Sea

Habitat: This species occurs from littoral to 60 m depth areas. It is often found in terebellid polychaete tubes.

Biology: Body of this scale-worm is of uniform width. Dorsal and ventral surfaces are smooth.

Prostomium is bilobed, and is usually with a small rounded peak on each lobe, and median and lateral antennae. Of its two pairs of eyes, anterior pair is near the front of the prostomium, and posterior pair is in front of the rear margin. First segment bears chaetae and a pair of dorsal and ventral tentacular cirri. A rounded facial tubercle is present. Scales overlap but do not cover the dorsal surface of the body. These scales occur only on the anterior region on segments 1, 3, 4 and 6, then alternately to the 22nd and then on segments 25, 28 and 31. Scales are rounded and their margins are smooth. Notopodia are small mounds with a ventral acicular ligule and several chaetae. Notopodial chaetae have rows of spines and blunt cleft tips. Neuropodia are well developed with an anterodorsal finger-shaped and a posterior rounded ligule. Coloration of the animal is dark red on the dorsal surface and yellow to reddish brown on the ventral surface. A dark pigmented streak is usually seen across the dorsal surface of each segment in the anterior region giving a striped appearance. An animal with 100 chaetigers may have a maximum length of 120 mm.

Bioluminescence: Nicol (1953) reported that when the anterior half of the body is cut across, the scales posterior to the cut become luminescent. Moraes *et al.* (2021) reported on its wavelength of emission maximum as 515 nm.

Robertianella synophthalma

Common name (s): Not known

Global distribution: Eastern Central Atlantic and the Mediterranean Sea

Habitat: This benthic, burrowing species is found on hard ground up to a depth of 2791 m. This worm is also found in association with the mud-dwelling sponge *Pheronema carpenteri*

Biology: Individuals of this species can grow up to 100 mm. Body of the worm consists of a head, a cylindrical segmented body and a tail. Its head consists of a prostomium and a peristomium and bears paired appendages.

Bioluminescence: The bioluminescence of this species is due to its elytra which showed brighter cells around the elytrophore scar. Its emission spectrum showed a maximum emission peak at 530 nm (Taboada *et al.*, 2020) and 511 nm (Verdes and Gruber,2017)

Subadyte pellucida

Common name (s): Not known

Global distribution: Indian, Pacific and North Atlantic Oceans

Habitat: It occurs from the intertidal zone to depths of 14- 800 m. It is also known to be a commensal organism, living on host brittle stars and sea stars

Image credit: Azote Library. (Applied for permission)

Biology: It is a short-bodied worm with about 40 segments and 15 pairs of elytra, which bear a marginal fringe of papillae. Its lateral antennae are positioned ventrally on the prostomium, directly beneath median antenna ceratophore. Notochaetae are about as thick as the neurochaetae, which also possess bidentate tips.

Bioluminescence: It is a bioluminescent species (Moraes *et al.*, 2021).

Family: Syllidae

Eusyllis blomstrandi

Image credit: Arne Nygren/Sjøfartsmuseet Akvariet Gøteborg, Wikimedia Commmons

Common name (s): Not reported

Global distribution: It is found on both sides of the North Atlantic, in the Arctic, and in the North Pacific.

Habitat: It is generally found in a mucus tube in shallow hard bottom areas with bryozoans, hydroids Laminaria and red algae from the sublittoral down to 1700 m deep.

Biology: Body of this species is thick but fragile.Prostomium is rounded quadrangular to semicircular with four eyes in a trapezoid arrangement and two ocelli. Palps are broad lobe-like, and are often larger than the prostomium.

Median antenna are about 2 times longer than lateral antennae. Lateral antennae are longer than palps and without annulation.

Tentacular cirri and dorsal cirri are without distinct annulation. Ventral cirri are present. Blades of compound chaetae are very short and bidentate. Posterior parapodia have additional upper capillary chaeta and lower bidentate simple chaeta. Pharynx has two rows of soft papillae anteriorly and is crowned by a dark, finely toothed chitinous ring (trepan). It also has a single large conical tooth. Color of the animal is yellow or orange and gills are with brown tips. Animals with 124 segments have a maximum length of 32 mm.

Bioluminescence: It has paired epidermal luminescent regions in each segment that produce light upon stimulation. Light emission is normally localized on a posterior group of segments, where it seems to be intracellular and can be turned on and off very rapidly (Verdes and Gruber,2017). Haddock *et al.* (2010) reported that its wavelengths of emission peak as 440-510 m. Brugler *et al.* (2018) state that this species emits light not only for reproductive purposes but also as defensive mechanisms. Further, they are able to detach their posterior luminescent part of their bodies to distract predators while the anterior end escapes.

Zorner and Fischer (2007) reported on the luminescent organs and spatial pattern of bioluminescent flashes of this polychaete. According to them, each of the trunk segments of this species is equipped with paired epidermal luminescent domains which luminesce upon mechanical or electrical stimulation. Light emission which is typically emitted in series of flashes is intracellular and it could rapidly turn on and off, and is heavily coordinated among the trunk segments. Further, in this species, the luminescent reaction usually involves only the posterior group of segments. In posterior fragments, all the segments participate in flashing luminescence. Amazinlgy posterior fragments are capable of luminescence weeks after fragmentation although they do not regenerate a head. Immediately upon fragmentation of the worm, the posterior fragment luminesces continuously for some seconds while the anterior part quickly stops the light emission.

Nudisyllis pulligera (= *Pionosyllis pulligera*)

Common name (s): Not known

Global distribution: North Atlantic to Mediterranean and English Channel

Habitat: It occurs in shallow water benthic habitats

Biology: Body of this species is short and is relatively broad. Prostomium is rounded quadrangular to semicircular with four eyes which are in a trapezoid arrangement. Palps are broad, and are slightly shorter than the prostomium. Median antenna is much longer than lateral antennae and is at least three times longer than prostomium. Lateral antennae are longer than palps.Dorsal and tentacular cirri are smooth or with unclear annulations. Dorsal tentacular cirri are as long as median antenna. Dorsal cirri are as tentacular cirri without annulation. Ventral cirri are long as parapodial lobes. Blades of compound chaetae are moderately long, unidentate or indistinctly bidentate. Additional simple chaetae may be present in posterior parapodia. Body coloration is whitish, with yellow cross-lines on anterior body or a median lateral line of purple spots. Body of the animal with 56 segments has a maximum length of 10 mm only.

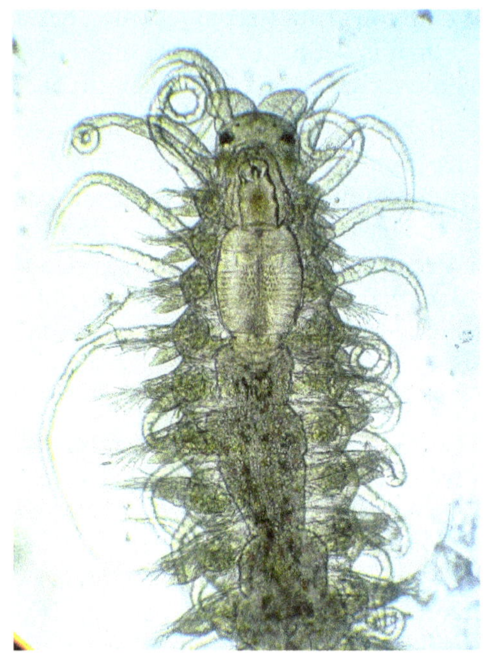

Image credit: Jo Langeneck, Wikimedia Commons

Bioluminescence: This species has been reported to display a clear phosphorescent display during swarming. Females swimming at the water surface display repeated periods of continuous light emission. Males also emit short flashes when they are quickly approaching the luminescent females (Dorrestein and Westheide, 2013).

Odontosyllis australiensis

Common name (s): Not reported

Global distribution: Japan, Australia (Western Australia, Queensland, New South Wales).

Habitat: It inhabits intertidal to shallow waters on dead coral, coarse sand and algae

Biology: In this species, prostomium is almost rounded and is with 4 eyes which are in open trapezoidal arrangement. Antennae which are originating near anterior margin are close to each other; median antenna is inserted slightly posterior to lateral ones; and lateral antennae are shorter than median one. Palps are shorter than prostomium and are ventrally folded. Peristomium is reduced dorsally, and is covered by occipital flap. Animals of 55 chaetigers have a maximum length of 6.6 mm.

Bioluminescence: It is a bioluminescent species (Verdes *et al.*, https://www.readcube.com/articles/10.1101/241570). Kotlobaya *et al.* (2019) reported that *Odontosyllis* bioluminescence reaction needs only oxygen, a luciferin, and a luciferase enzyme to emit its greenish-blue light. During the enzymatic luminescent reaction, luciferin is converted into an oxyluciferin (light emitter), (Kotlobaya *et al.*, 2019)

Odontosyllis ctenosoma

Common name (s): Not reported

Global distribution: Mediterranean and eastern Atlantic areas

Habitat: It is inhabiting photophilic infra-littoral hard substrates and *Posidonia oceanica* rhizomes.

Biology: It is a relatively large, dioecious, epigamic syllid. Maximum lengthnof the animal with 100 chaetigers is 20 mm.

Bioluminescence: It is a bioluminescent worm (Verdes *et al.*, https://www.readcube.com/articles/10.1101/241570)

Odontosyllis enopla

Common name (s): Bermuda glow worm, Bermuda fireworm

Global distribution: Western Central Atlantic and Eastern Central Pacific

Habitat: This benthic species is found in in soft substrate, shallow areas and it has a depth range of 10 - 189 m. This worm has been reported to live normally in a mucous tube on the shallow seabed.

Biology: It has a body consisting of multiple segments each with a pair of parapodia. Head has two pairs of eyes at the sides these eyes possess a lens, photoreceptor cells, a retina and pigment granules. Each pair of eyes is oriented in a different plane. While female worms grow to 20 mm males grow to 12 mm.

Bioluminescence: In this species, females show a bright continuous glow while swimming rapidly in small circles. Subsequently, the males swim directly to the females, emitting flashes of light. They both rotate together releasing gametes in the water column. Females normally glow for 3–8 s, while males give 2–3 quick flashes of light when they are approaching a female. It is also reported that the display lasts roughly half an hour, beginning around 55 min after sunset, each month after a full moon. In this species, the components of the bioluminescence system include Luciferin/luciferase/Mg2+/cyanide/O2 and the wavelength of the emission peak is 507 nm (Verdes and Gruber, 2017)

Brugler *et al.* (2018) reported that all the fire worms are believed to secrete mucus which emits bluish-green light with a luciferin-luciferase reaction.

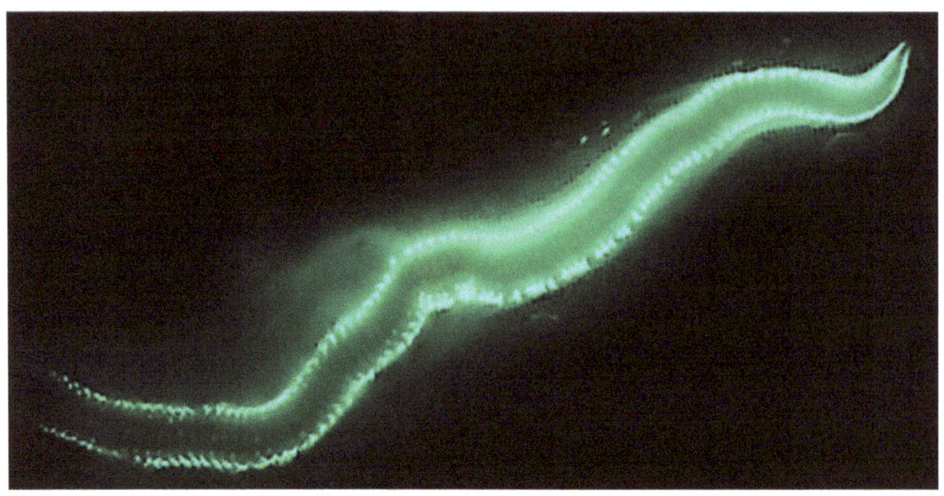

Bioluminescent Display of *Odontosyllis enopla*

Image credit: James B. Wood, Reproduced with permission

Odontosyllis luminosa

Common name (s): Japanese syllid polychaete

Global distribution: Western Central Atlantic: Central America to northern Brazil.

Habitat: This benthic species inhabits soft-sediment habitats at depths beyond 45 m.

Biology: This species lacks the defined patterns of the dorsal pigmentation but it has much longer antennae and cirri throughout. Further it has characteristically modified chaetae.

Bioluminescence: Verdes and Gruber, (2017) reported that both *Odontosyllis enopla* and *Odontosyllis luminosa* possess more or less similar swarming and bioluminescent displays which happen during the second night after a full moon, around 75 min and 55–65 min after sunset, respectively, especially during the summer months

Gaston and Hall (2000) reported that the female glowworms of this species leave the sediments, swim toward the water surface, and release a bioluminescent egg mass which causes a bright green glow near the water's surface. This female's luminescence attracts the males, which also simultaneously glow briefly. Reproduction of this species peaks during summer, when the water's surface is brilliantly lit with females for 10–15 min on the first few evenings following a full moon. Spent females apparently survive to spawn again. They then return to the sediments to build new tubes after spawning

Odontosyllis octodentata

Common name (s): Not known

Global distribution: Western Central Atlantic: Central America to northern Brazil.

Habitat: This benthic species is a burrowing form.

Biology: Individuals of this species can grow to 50 mm. No other information

Bioluminescence: As for *Odontosyllis luminosa*

Odontosyllis phosphorea

Image credit: Peter J. Bryant. (Applied for permission)

Common name (s): Fireworm

Global distribution: Pacific coast of North and Central America

Habitat: This organism normally lives in a tube on the seabed from the intertidal zone down to the continental shelf. This species is typically found among seaweed growing on rocks and among seagrasses such as *Zostera.*

Biology: It is a small worm from 20 to 30 mm long and 1 mm in dia. when it is fully grown. Its elongated body is composed of many segments and each one is bearing a pair of parapodia. Head has two pairs of eyes, a nuchal hood which covers the back of the prostomium. Its parapodia in the central part of the body are slender and tapering. Upper surface of the worm is dark with yellowish transverse bands

Bioluminescence: It becomes bioluminescent when it rises to the surface of the sea during breeding season during the last quarter or beginning of first quarter of the moon, about half an hour after sunset. Both the bioluminescent display and spawning are known to last around 30 min. Unlike other *Odontosyllis* species, in this species, the first flashes of light are usually from males. The females appear shortly after, swimming in circles, flashing and secreting a bright green luminous mucus along with the gametes (Verdes and Gruber, 2017)

Deheyn and Latz (2009) reported that this species produces brilliant displays of green bioluminescence during mating swarms. In laboratory conditions, it was found that the light emission appeared as an intense glow after stimulation with potassium chloride, and the emission was associated with secreted mucus. The mucus was found to be viscous, blue in color, and displayed a long-lasting glow that was however, greatly intensified by the addition of ammonium persulfate or peroxidase. Further, the emission spectrum of the mucus-associated bioluminescence was found to be unimodal, with a maximum emission in the green spectrum between the wavelengths 494 and 504nm. It was also found that the luminous system was functional at temperatures below –20°C and was almost degraded above 40°C. While the mixing of hot and cold extracts of the mucus did not result in light emission, mucus samples exposed to oxygen depletion by bubbling with argon or nitrogen were able to produce intense bioluminescence. According to these authors, these results suggested that bioluminescence from the mucus of this species may involve a photoprotein rather than a luciferin–luciferase reaction.

Odontosyllis polycera

Common name (s): Not reported

Global distribution: Pacific Ocean; Angola, Namibia, South Africa, USA (Southern California), Panama, Indo-Pacific, New Zealand, Australia

Habitat: This benthic species is occurring in sand, mud, algae, calcareous substrata, bryozoans, sponges, from intertidal to a depth of about 90 m.

Biology: Body of this species is robust broad anteriorly, with numerous segments. Prostomium is oval with 2 pairs of large eyes in open trapezoidal arrangement. Median antenna is longer than combined length of prostomium and palps. Lateral antennae are inserted close to median antenna, on anterior margin of prostomium. Palps are divergent, ventrally folded, and are fused basally. Dorsal tentacular cirri are slightly longer than antennae and ventral tentacular cirri are shorter. Peristomium is reduced dorsally and is covered by large, long occipital flap, that also covers most of prostomium. An animal with 116 chaetigers may have a maximum length of 25 mm.

Bioluminescence: This species becomes bioluminescent one month prior to spawning. Further, males are not only stimulated by the bioluminescent display of the female, but also by the beam of a flashlight (Wikipedia)

Odontosyllis undecimdonta

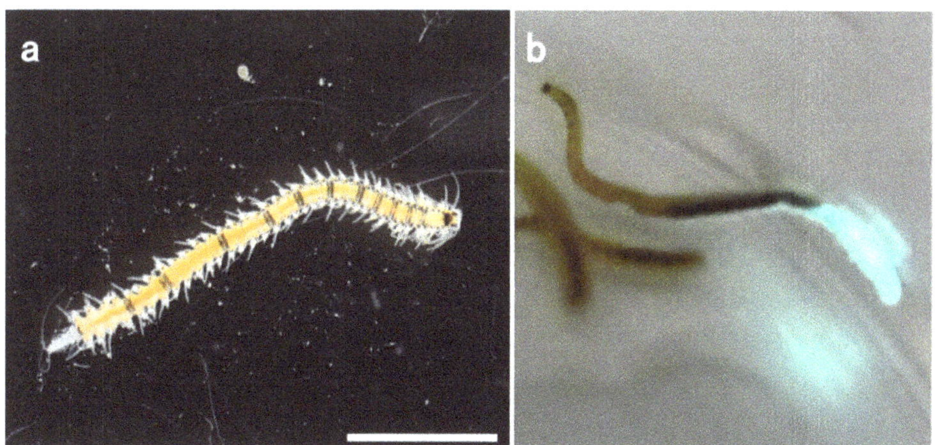

Animal and its Bioluminescence

Image credit: Yasuo Mitani1, RieYasuno2, Minato Isaka1,3, Nobutaka Mitsuda, Ryo Futahashi, Yoichi Kamagata and YoshihiroOhmiya. (Applied for permission)

Common name (s): Japanese fireworm

Global distribution: North West Pacific Ocean (Japan).

Habitat: This benthic species occurs in soft sediments

Biology: Dorsum of this species is distinctly arched and is ventrally flattened. Prostomium is wider than long, rectangular to oval and is with two pairs of large, red eyes. Of these eyes, anterior ones are slightly larger. Palps are broad, and are fused at bases. These palps are similar in length to prostomium and are ventrally folded. Animal with 71 segments showed a length of 13 mm.

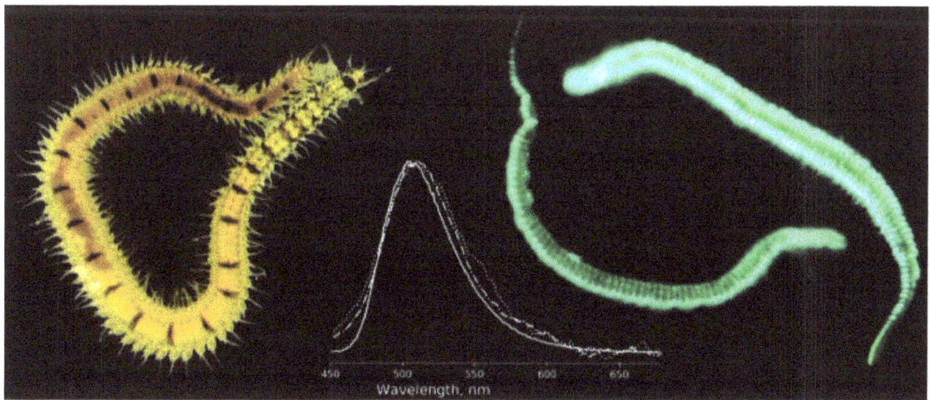

Animal and its Bioluminescence

Image credit: Ilia V. Yampolsky and Yuichi Oba. (Applied for permission)

Bioluminescence: In this species, its biosynthesized and secreted *Odontosyllis* luciferase takes part in its bioluminescence. It has been reported to emit bright blue-green bioluminescence and it is known to display a spectacular bioluminescence, correlated with the lunar cycle (Kotlobaya *et al.,* 2019)

Brugler *et al.* (2018) reported that the luciferase of this species is "evolutionarily unique" as no identifiable homologous proteins has so far been found

Order: Terebellida; Family: Acrocirridae

Swima bombiviridis

Image credit: Casey Dunn, Reproduced with permission

Common name (s): Green bomber worm, bombardier worm.

Global distribution: Monterey Bay, California.

Habitat: This entirely holopelagic species lives in the ocean sediment at the depths of 1–450 m. above the sea floor.

Biology: The characteristic features of this transparent worm are the presence of four pairs of elliptical, transformed segmental branchiae on anterior segments and a gelatinous sheath. It can grow over 30mm in length and 5 mm in width.

Bioluminescence: The elliptical branchiae (modified bioluminescent gills) that it employs to drop 1mm long bioluminescent, green 'bombs' that luminesce for several seconds after they have been discarded (https://en.wikipedia.org/wiki/Swima_bombiviridis)

Verdes and Gruber(2017) reported that the bioluminescent structures of this species are nothing but the four pairs of modified branchiae that immediately produce green light when autotomized. It is also known that mechanical stimulation of any part of the body results in the detachment of a bomb

Bioluminescence in *Swima bombiviridis*

Image credit: Karen J. Osborn and Steven H.D. Haddock (Applied for permission)

Swima fulgida

Common name (s): Shining bomber worm

Global distribution: Northeastern Pacific Ocean; Central California coast

Habitat: This benthopelagic species is swimming in the waters just above the seafloor at depths of 30 –3,600 m. In this species, animals are known to hang horizontally in the water column with its dorsal surface uppermost and the palps hanging forward and downward over the buccal organ

Biology: A darkly pigment anterior gut and buccal organ are the characteristic features of this species. This transparent worm bears four pairs of elliptical, transformed segmental branchiae on anterior segments; a gelatinous sheath which is penetrated throughout by narrow clavate papillae; simple noto- and neurochaetae; and three achaetous anterior segments supporting ellipsoid, bioluminescence-producing, derived branchiae that are less than 1.2 mm in length.

Image credit: K. J.Osborn, S.H.D.Haddock and G.W.Rouse (Applied for permission)

Bioluminescence: A typical bomber worm has eight sacs of bioluminescent fluid just behind its head. When disturbed, the worm releases a "bomb" that bursts with a green glow. Amazingy,it is capable of regenerating its lost bombs (Gouveneaux (2016;https://www.mbari.org/products/creature-feature/bomber-worm/)

Swima tawitawiensis

Image credit: Karen J. Osborn and Steven H.D. Haddock. (Applied for permission)

Common name (s): Orange bomber

Global distribution: Northeastern Pacific Ocean

Habitat: This benthopelagic species swims in the waters a few meters (about 30 m) above the seafloor and occurs at depths of 2,700–3,600 m

Biology: Diagnosis: It has nuchal organs, each with medial end curved into U-shape. There are three equally long, subulate branchiae just posterior to nuchal organs. Of them one is medial and two are lateral in position. Notochaetae are broad and flattened with a fine, spinous tip.

Bioluminescence: It bioluminescent and produce light by chemical reactions within its body. This transparent worm bears four pairs of elliptical, transformed segmental branchiae on anterior segments that produce green bioluminescence when autotomised. These luminescent capsules (bombs), which are not all released at once when the organisms are manually stimulated, are able to regenerate. (https://sumyuworld.wordpress.com/2012/02/06/deep-sea-beasties/)

Family: Cirratulidae

Aphelochaeta multibranchis (= *Tharyx multibranchis; Heterocirrus multibranchis*)

Common name (s): Not reported

Global distribution: Northeast Atlantic up to Mediterranean, North Sea.

Habitat: It is found from eulittoral to middle sublittoral, down to 128 m deep.

Biology: Body of this species is cylindrical and its mid-body is slightly swollen.Prostomium is conical and is fused to the anterior segments, with 2 small black eyes. Anterior three achaetous segments get dorsally fused together. One pair of long, thick palps is seen on the last achaetous segment. The filamentous gills which are present on all but the very last segments are found inserted dorsally starting at the 1st chaetiger. The distance between gills and notopodia is longer than between noto- and neuropodia.

Chaetae are smooth capillaries. Dorsal chaetae very fine and ventral chaetae especially in the posterior body half are slightly shorter, thicker and slightly curved. Colour of the body is yellowish-green or light pink. A worm with 65 segments had a maximum length of 10 mm.

Bioluminescence: In this species, its epidermal cells opening between the chaetae of each segment emit light (Kin *et al.*, 2021).

Caulleriella bioculata (= *Heterocirrus bioculata*)

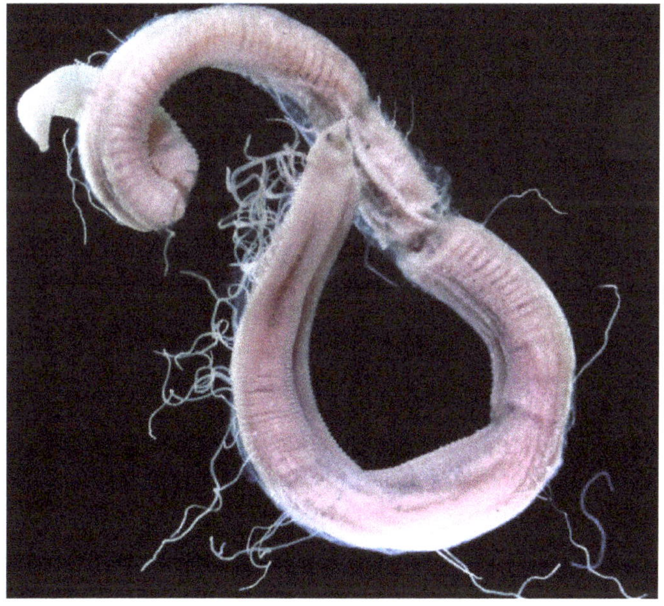

***Caulleriella* sp.**

Image credit: EOL (CC)

Common name (s): Not known

Global distribution: Indo-Pacific, Northeast Atlantic and the Mediterranean.

Habitat: It inhabits soft bottoms

Biology: Prostomium of this species is conical, with two eyes on the posterior half. It also has two dorso-lateral lobes. Anterior two achaetous segments get dorsally fused. Achaetous segments are longer than chaetigers. One pair of long thick palps is seen on the first chaetiger. Slender filamentous gills, starting at first chaetiger, are present predominantly on anterior body half. Chaetae are capillary and are very finely serrated. Colour of the animal is greenish-yellow with yellowish or brown pigmentation. A worm with 140 segments may have a maximum length of 40 mm.

Bioluminescence: It emits yellow-green light when it is stimulated by freshwater (Kin *et al.*, 2021)

Caulleriella fragilis (= *Cirratulus fragilis; Cirrhinereis phosphorea*)

Common name (s): Red thread worm

Global distribution: NE American east coast.

Habitat: This benthic species is found under intertidal rocks

Biology: Body of this species is cylindrical, and is narrowed towards the extremities. Body coloration is reddish orange and posteriorly it is greenish. Mouth is inferior and circular; and upper lip is conical. There are two eyes. Cirri are numerous and are orange colored. First pair is commencing at the second setigerous segment and is most robust. Setae are seen in two rows and are simple. It can grow up to a length of 100 mm.

Bioluminescence: In this species, both sexually mature and atokous (not sexually mature) individuals have been reported to produce a bluish light when disturbed- (Verdes and Gruber, 2017; Kin *et al.*, 2021)

Caulleriella parva

Common name (s): Not known

Global distribution: It is known only from Helgoland (a treeless island in the North Sea)

Habitat: Not reported

Biology: Prostomium of this species is bluntly conical, with two black eyes. Second achaetous segment is the longest and is with a dorsal lobe which is slightly covering the third segment. One pair of palps is seen starting on the first chaetiger. Filamentous gills start at the first chaetiger just behind the palps and they are posteriorly becoming smaller and fewer.

Anterior parapodia only are with capillary chaetae. Additional bidentate chaetae are present from third neuropodium and from 7-11th notopodium.

Colour of the animal is yellowish-green, especially the prostomium and tips of palps and gills. A worm with 36 segments has a maximum length of 34 mm.

Bioluminescence: It has been reported to emit bluish light when it is pinched or irritated with freshwater (Kin *et al.*, 2021; Verdes and Gruber, 2017)

Chaetozone caputesocis (= *Caulleriella caputesocis*)

***Chaetozone* sp. (Anterior end)**

Image credit: James A.Blake and Nicolas Lavesque (CC)

Common name (s): Not known

Global distribution: Northeast Atlantic, Mediterranean and Western Indian Ocean: from Germany to France, east to Red Sea.

Habitat: This benthic species has a depth range of 30 - 66 m

Biology: Body of this species is cylindrical and slender. Prostomium is bluntly conical, with two relatively large eyes on the posterior half of the prostomium. Eyes are surrounded by small pigment patches. Prostomium has two dorso-lateral lobes. Anterior two achaetous segments get dorsally fused. Its long, thick palps are inserted on the last achaetous segment. Filamentous gills start at first chaetiger. Gills are absent on posterior 14-30 segments.

Chaetae are capillary and smooth in anterior segments; and smooth and finely serrated in posterior notopodia. Color of the animal is yellowish-brown, partly with pigmentation near the eyes. A worm with 95 segments may have a maximum length of 17 mm.

Bioluminescence: It has been observed to emit a greenish light from photophores found mainly in the epidermis of abdominal segments (Verdes and Gruber, 2017; Kin *et al.*, 2021)

Dodecaceria saxicola (= *Heterocirrus saxicola*)

*Common name (s):*Not known

Global distribution: Mediterranean Sea.

Habitat: It is seen subtidally in gravel and stony ground

Dodecaceria sp.

Image credit: Arne Nygren/Sjøfartsmuseet Akvariet Gøteborg, Wikimedia Commons

Biology: The species of this genus have acicular chaetae swhich are poon-shaped, with or without conical projection. Palps and gills are thick and fewer than 8 pairs. Prostomium is broadly rounded, with large nuchal organs.

Bioluminescence: It has been reported to emit green light (Kin *et al.*, 2021)

Tharyx sp.

Common name (s): Not known

Global distribution: Japan

Habitat: The species of this genus occur at depths of about 380 m.

Biology: Body coloration of this species is yellowish. Prostomium is conical and eyespots are absent. Nuchal organs are not seen. Peristomium is elongate with a pair of dorsal tentacles. Pairs of branchiae are seen on dorsolateral side of body; and first pair of branchiae on achaetous segment are located posterior to peristomium. Parapodia are biramous, with 4–5 capillary notochaetae and 2–3 capillary neurochaetae/2–3 neurospines.

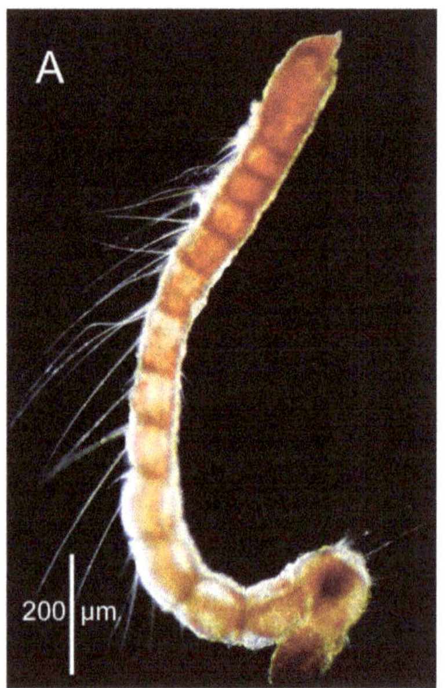

Image credit: Blake, J.A. Wikimedia commons

Bioluminescence: The bioluminescence of the Tharyx specimens has been reported to be greenish; and they emitted the light when they are disturbed/pinched (Kin *et al.*, 2021).

Family: Flabelligeridae

Poeobius meseres

Common name (s): Not known

Global distribution: This subtropical species occurs in northern Pacific from Japan to the Gulf of California

Habitat: This holopelagic and swimming polychaete is typically found in midwater depths of 25-3975 m

Biology: Body of this species is largely gelatinous with a thick mucus sheath and segmentation is not clearly visible. Anterior end bears retractile pale green tentacles which consist of a pair of grooved palps and branchiae. It has 11 poorly defined segments and it grows up to 27 mm in length.

Bioluminescence: This species has been reported to generate light along the whole length of the body and its light emission peaks ranged from 493 to 497 nm (Francis *et al.*, 2016). Verdes and Gruber (2017) reported that this species produces flashes of light as spots near the bristle junctions

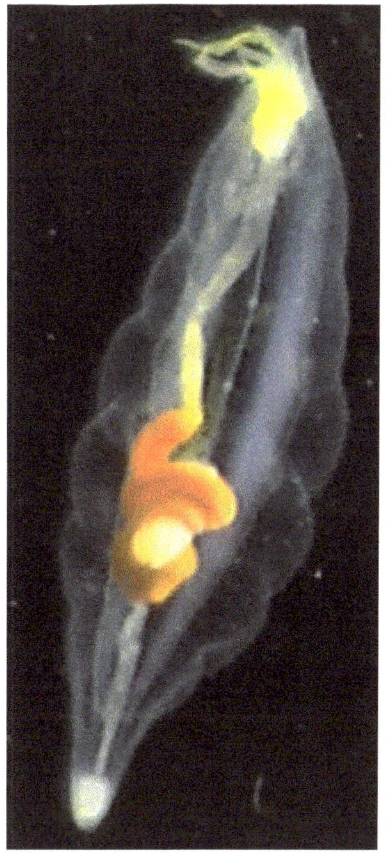

Image credit: Charlotte Seid, Reproduced with permission

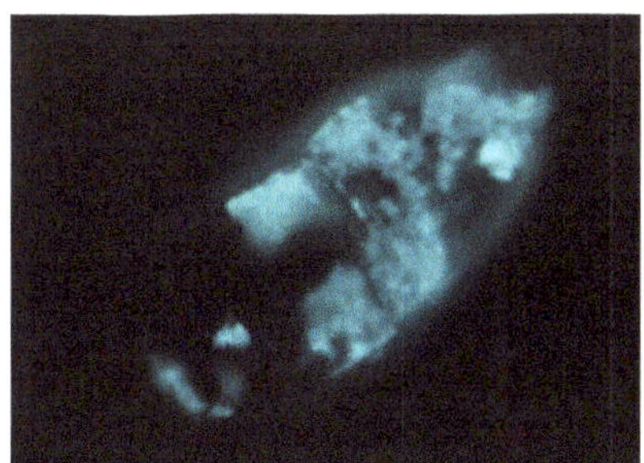

Left: Bioluminescence; Right: Animal

Image credit: Steve Haddock, Reproduced with permission

Flota flabelligera

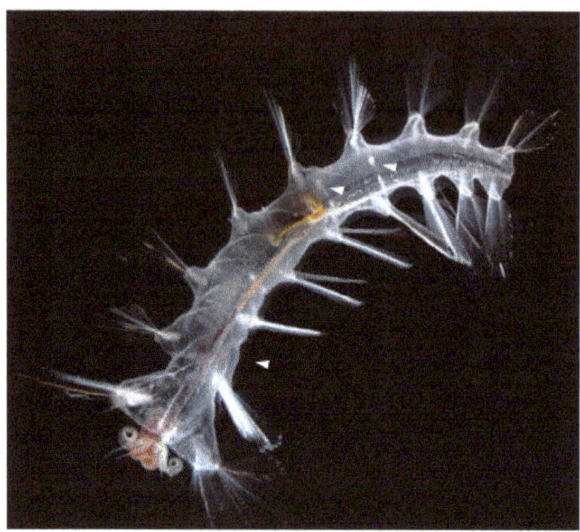

***Flota* sp.**

Image credit: Karen J. Osborn and Greg W. Rouse (Applied for permission)

Common name (s): Bristle worm

Global distribution: South Atlantic Ocean; southwestern Chile and off Cape Horn

Habitat: This benthic facultatively mobile species occurs at a depth of about 3800 m

Biology: Body of this bristle worm consists of a head, a cylindrical segmented body and a tail. Head consists of a prostomium (part in front of the mouth opening) and a peristomium (part around the mouth) and bears paired appendages (palps, antennae and cirri).Individuals of this species can grow up to 100 mm.

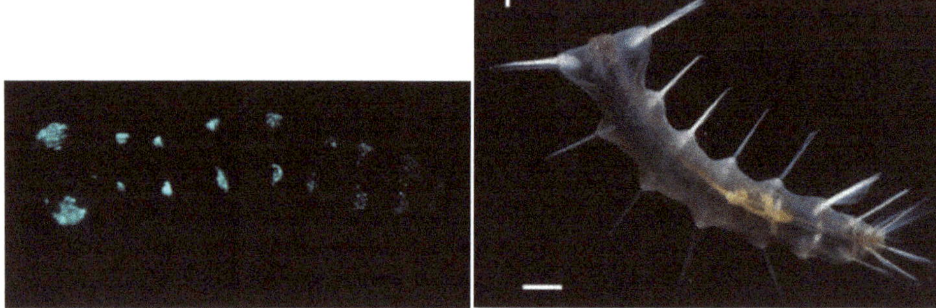

Left: Bioluminescence; Right: Animal

Image credit: Steve Haddock, Reproduced with permission

Bioluminescence: This species has been reported to generate light along the whole length of the body (Francis *et al.*, 2016). Verdes and Gruber(2017) observed that it produces flashes of light as spots near the bristle junctions

Family: Terebellidae

Polycirrus aurantiacus

Image credit: Deviantart.com. (Applied for permission)

Common name (s): Not reported

Global distribution: Northeast Atlantic and the Mediterranean Sea: Germany.

Habitat: It occurs on soft bottoms or among algae, serpulids, hydroids or dead shell. Its tube is mucous and is not lasting but it is phosphorescent.

Biology: Body of this species is very short; dorsum is convex; and ventral side is flattened. There are numerous tentacles which are composed of very long cylindrical ones, and short, distally thickened ones. Tentacular ridge is convex. One unpaired, long, cushion-like, and 8-11 paired, distinct ventral shields are present all distinct. Notopodial chaetae are very finely feathery, long and short and they are all starting at segment 2 and in at least 30-40 segments. Notopodia are cylindrical. In life, its body coloration is dark orange and tentacles are orange, bluish or violet phosphorescence. A worm with 120 segments may have a maximum length of 100 mm.

Bioluminescence: This species has been reported to produce a bright violet-blue luminous secretion in the tip of its tentacles and its bioluminescence is of extracellular nature (https://www.deviantart.com/monsieur-le-gris/art/Polycirrus-aurantiacus-179329719)

Polycirrus caliendrum

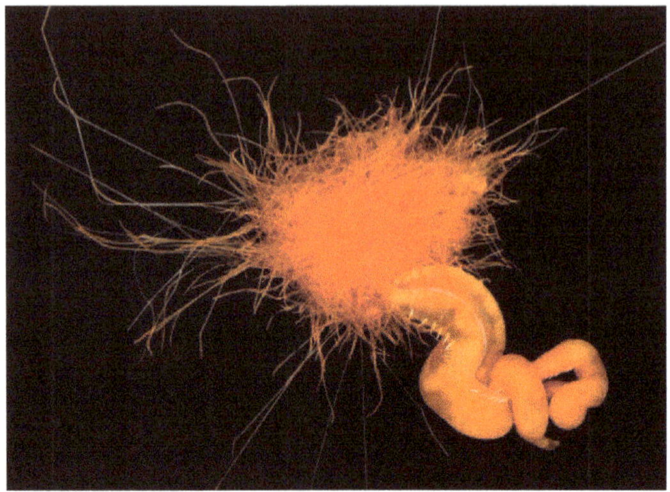

The Head End of *Polycirrus caliendrum* is Surrounded by a Seething Mass of Tentacles.

Image credit: NPL Second Nature (Applied for permission)

Common name (s): Not reported

Global distribution: Eastern Pacific and the Mediterranean Sea: Mexico, Peru and Galapagos Islands.

Habitat: It occurs on soft bottoms or among algae, serpulids, hydroids or dead shells. Its tube is mucous and is not lasting but is phosphorescent

Biology: Body of the animal is soft. Tentacular ridge and upper lip are pleated. Tentacles are of two types, *viz.* long cylindrical ones and short, distally thickened ones. One unpaired and 8-11 paired ventral shields are seen. Notopodial chaetae are long and short and they are starting at segment 3. Its 30-70 segments have cylindrical notopodia with chaetae. In life, animal is orange to yellow in coloration. Tentacles are yellowish or more or less pale. A worm with 90 segments has a maximum length of 100 mm.

Bioluminescence: Huber *et al.* (1989) reported that it has extracellular nature of bioluminescence.

Polycirrus perplexus

Common name (s): Not reported

Global distribution: USA, California, Monterey Bay off Santa Cruz Island,

Habitat: This benthic species occurs at a depth of about 220 m.

Biology: In this species, its dorsum is anteriorly tessellated. Anteriorly venter has mid-ventral groove and discrete ventro-lateral pads which are incised, tessellated and are extending from segment 3 to 14. Mid-ventral groove is seen from segment 4. Body is greyish in colour and a worm with 75 segments is 30 mm long and 3 mm wide.

Bioluminescence: This species is known for its blue bioluminescence. It is capable of producing very rapid flashes and its lambda maximum wavelength is unusually short *i.e.* 445 nm which is due to its coastal, benthic habitat. Further, bioluminescence in this species is not secretory ((Huber *et al.*, 1989; Verdes and Gruber,2017).

Image credit: Michael E. Huber, A. Charles Arneson and E. A. Widder. (Applied for permission)

Thelepus cincinnatus

Common name (s): Not reported

Global distribution: Northeast Atlantic, the Mediterranean Sea, Arctic and Antarctic.

Image credit: WoRMS (CC)

Habitat: It occurs on most kinds of substrata which range from mud through sand to solid rocks. It is especially common and abundant on coarse sand with shells. It is an euhaline species and it is occasionally polyhaline. This benthic species may have a depth range of 7 - 1950 m. Its tube is irregularly curved, and is often very long. It is reported to be a tough, transparent layer of secretion which is often firmly glued to stones or large shells and the free surface is encrusted with coarse sand grains and shell fragments.

Biology: Body of this species is long and its anterior part is cylindrical and posterior part is tapering.There are numerous tentacles which are very long and thick. Upper lip is low, broad and thick. Many eyespots are seen. There are two series of gills (on segments 2 and 3), and each series is consisting of a number of free filaments arranged in transverse rows on dorsum. These filaments are known to contract spirally. Notopodial chaetae are smooth, and are of two types *viz.* long straight ones, and short distally curved ones. In life, body coloration is pink, orange or brownish with lighter ventral and lateral areas. Gills are red and tentacles are pale pink, orange or brownish with red spots. An animal with 100 segments may have a maximum total length of 200 mm.

Bioluminescence: In this species its bioluminescence is due to its luminous secretion (Verdes and Gruber,2017)

Thelepus japonicus

Image credit: EOL (CC/NC)

Common name (s): Not reported

Global distribution: British Columbia, China, Japan, Korea, Kuwait, Bay of Biscay (Arcachon Bay), English Channel

Habitat: This species occurs in oyster reefs or in oyster farm

Biology: In this species, prostomium is present at base of upper lip; eyespots are seen in thin and continuous dark band; and buccal tentacles are long, and deeply grooved. Peristomium is continuing dorsally as narrow annulation. Individuals are about 67–150 mm long and 5.0– 6.5 mm wide.

Bioluminescence: This species has been reported to produce light using a novel system which differs from that in other known luminescent annelids. The extracts of this species emitted blue-green light at λmax 508 nm following the addition of divalent cations like Fe2+. It is also worth mentioning here that addition of ATP, H2O2 or coelenterazine did not enhance luminescent activity (Kin *et al.*, 2019)

3.6. Crustaceans

3.6.1. Amphipods

Subphylum Crustacea; Class: Malacostraca; Order: Amphipoda

Among the swimming amphipods, the epibenthic gammarids *viz.* the species of the genus Cyphocaris which are found throughout the world's oceans from 90 m to thousands of meters depth. are known for their bioluminescence and they have been reported to emit orange light (595 nm) through an unknown mechanism (Haddock *et al.*, 2010). This phenomenon is thought to be related to their feeding strategies to capture prey (Hughes and Lowry,2015)

3.6.1.1. Hyperiid Amphipod

Megalanceola stephenseni (= *Megalanceola terra-novae*)

Common name (s): Not known

Global distribution: North Atlantic, near the Azores; Newfoundland, Bermuda and Southern Ocean

Habitat: It is usually found in deeper water (about 1000 m)

Biology: Head of the individuals is short, with a very small rostrum. Bbody is massive, with a bulging or flattened pereon. Integument is thick. Eyes are narrow, reniform, and relatively large. Mature females grow up to 73 mm, and males are up to 48 mm

Bioluminescence: Bowlby *et al.* (1991) reported that this species is known for its secretory luminescence.

3.6.1.2. Gammarid Amphipods

Cyphocaris anonyx

Common name (s): Not known

Global distribution: Indian, Pacific,and Atlantic Oceans; Sub- Antarctic; south-west Greenland

Habitat: It has a depth range of 0–2523 m

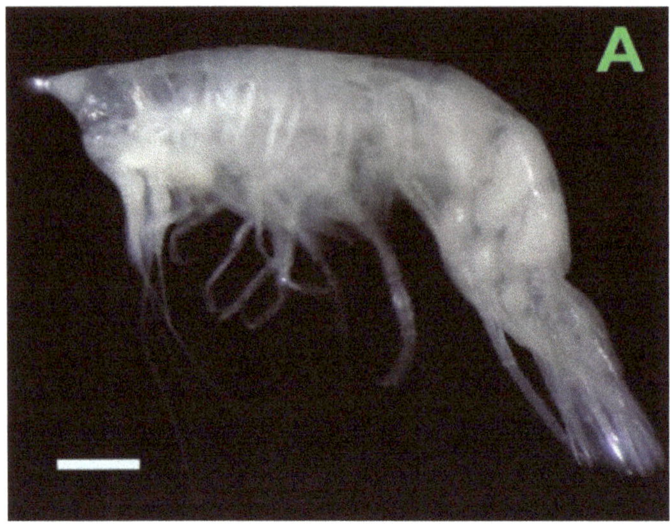

Image credit: Marco Violante-Huerta, Reproduced with permission

Biology: Body of this species is smooth and is without setae. Head is narrow, and is much deeper than long. Eyes are well-defined,pear-shaped, and bear few ommatidia. Rostrum is absent

Bioluminescence: It is a bioluminescent species (Bowlby *et al.*, 1991). No other information is available

Cyphocaris challengeri

Image credit: Ecology WA, Wikimedia Commons

Common name (s): Not reported

Global distribution: Pacific, Atlantic, and Indian Oceans

Habitat: It is a mesopelagic species having a depth range of 0–5987 m

Biology: Individuals of this species possess a maximum length of 13 mm. Eyes are large and reniform. In live animals "swords" projecting laterally. Live animals are pale-brown in coloration.

Bioluminescnece: This species has been reported to possess autogenous luminescence (Bowlby *et al.*, 1991)

Cyphocaris faurei

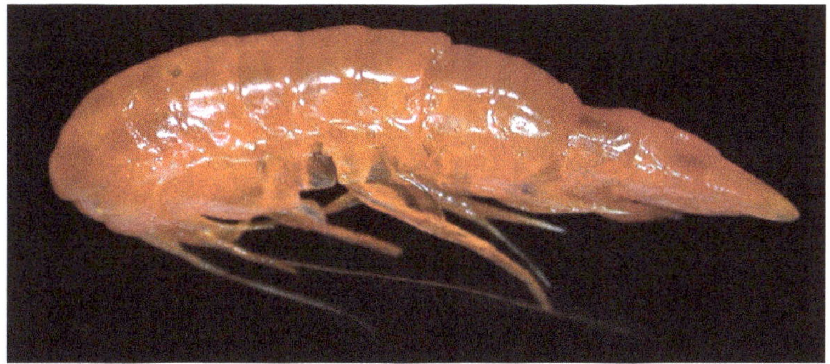

Image credit: Lauren E.Hughes and James K.Lowry. (Applied for permission)

Sites producing luminescence are indicated by stars.

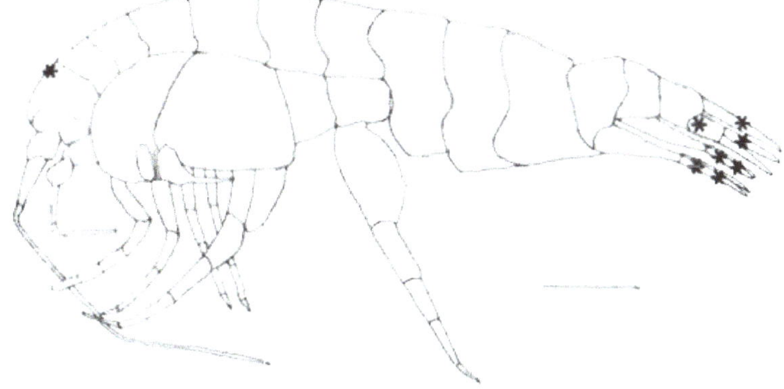

Image credit: Edith Widder, Reproduced with permission

Common name (s): Not reported

Global distribution: Indian, Pacific, Atlantic and Southern oceans

Habitat: It has a depth range of 175–900 m

Biology: In this species, eyes are pear-like and are broadened ventrally. While males grow to a maximum length of 30 mm, females are of 33 mm. Live animals are with red-rose coloration

Bioluminescence: In this species, bioluminescence appeared as a secretion through integumentary pores on the telson and uropods, and as a glow from a single location on the cephalothorax. Individuals produce light peaking in the blue-green region of the spectrum, averaging 478 nm. An additional second peak at 595 nm was also observed in some specimens (Bowlby *et al.*, 1991).

Cyphocaris richardi

Image credit: SCAR Antarctic Biodiversity Portal; Belgian Biodiversity Platform (Applied for permission)

Common name (s): Not known

Global distribution: Indian, Pacific, Atlantic and Southern oceans

Habitat: It has a depth range of 0–6145 m.

Biology: In this species, eyes are small, oval and fint. Live animals possess red-orange coloration. While males grow to a length of 40 mm females are of 56 mm in length

Bioluminescence: In this species, luminescence was induced by repetitive mechanical stimulation. It emitted a complex flash of longer duration and less quantum emission than that in *Cyphocaris faurei* (Bowlby *et al.*, 1991)

Pontogammarus maeoticus

Common name (s): Not known

Global distribution: It is endemic to the Caspian, Black and Azov Seas

Habitat: It is a hallow-water, sand- burrowing species,

Image credit: iNaturalist. (Applied for permission)

Biology: Body colour of this species is usually pale yellow. Both pairs of antennae are short and are of more or less the same length. It has an average length of 11 mm.

Bioluminescence: This species has been reported to emit constant uniform greenish light by the entire body, coupled with high mortality. It is also suggested that its light emission may be due to bacteriogenic bioluminescence. (Copila'-Ciocianu and Pop,2020)

3.6.2. Pycnogonids

Pycnogonids commonly known as "sea spiders" are benthic organisms inhabiting areas from intertidal zones to depths over 6,000 m. However, majority of them live on seaweed and corals. There are 4-6 pairs of walking legs and 2 pairs of dorsally located simple eyes which may be missing in deep-sea species. These species either walk along the bottom or swim just above it. The most distinctive feature of pycnogonids is the presence of a proboscis (a drinking straw) that allows them to suck out fluids from soft-bodied invertebrates such as corals and anemones. There are over 1300 species of deep sea spiders which are found in every ocean throughout the world. These organisms may grow up to 90cm

Family : Colossendeidae (Giant sea spiders)

The genus Colossendeis of this family includes the largest pycnogonids, with leg spans of about 40–50 cm. This genus contains bioluminescent species which are found in deep-sea living at depths of 2,200-4,000 m (https://en.wikipedia.org/wiki/Colossendeis; Haddock, *et al.*, 2010; https://eol.org/pages/14813/articles)

Colossendeis megalonyx

Common name (s): Southern Ocean giant sea spider

Global distribution: Antarctica, South America, Africa and Madagascar

***Colossendeis* sp.**

Image credit: NOAA Ocean Exploration and amp; Research from USA (Applied for permission)

Habitat: It is found both in the coastal waters and deep seas up to a depth of 4900 m.

Biology: The presence of eight long and lanky legs make these species easy to move along the deep seafloor. Its leg span of 25 cm is more or less equal to that of the world's largest land spiders like the Goliath Bird-Eating tarantula of South America. It is a carnivore using its elongate, tube-like proboscis to slurp up its prey which includes worms, jellyfish and sponges. Curiously, these organisms don't really have a body and have their organs in their legs. These creatures are therefore called the pantopoda – meaning "all legs" – because of their bizarre anatomy. Further, these sea spiders don't need to bother with a respiratory system and simple diffusion can deliver gases to all of its tissues

Bioluminescence: Certain species of the genus *Colossendeis* have been reported to be bioluminescent (Herring,1987; https://en.wikipedia.org/wiki/Colossendeis). However, no other information is available

3.6.3. Ostracods

Subphylum Crustacea; Class: Ostracoda; Order: Myodocopida; Family: Cypridinidae

The ostracods (sea fireflies, mussel shrimp or seed shrimp).have been reported to make dramatic displays of luminescence at dusk in the shallow reefs of the tropical Atlantic. Similarly, about 60 species of luminescent cypridinids that engage in luminescent courtship behavior have been discovered in the Caribbean (Haddock *et al.*, 2010). These ostracodes produce their own luciferin and luciferase in separate glands in the light organ, and secrete these molecules from a series of nozzles on the upper lip. Upon contact with the sea water, a bright blue luminescence is produced. Among the ostracods, the cypridinids

(Cypridinidae) provide the best marine examples as 60 per cent of the known 250 species are bioluminescent. Cohen and Morin (2017) reported that chemically different bioluminescence occurs in two groups of ostracods. While the bioluminescence is produced within carapace glands of certain Halocypridina, it is produced and extruded from the upper lips of many cypridinid Myodocopida. However, this bioluminescence serves as an antipredatory behavior in both the groups. It is hypothesized that courtship displays of ostracods evolved from this pre-existing anti-predatory function of the luminescence because more basal species use luminescence only for deterring predators and all signaling species also use their luminescence to deter predators. Males of many species of bioluminescent cypridinid ostracods produce nightly bioluminescent displays to attract females. These displays are species specific, and usually begin about an hour postsunset (Kates, 2011). Apart from serving as an important component of the food chain in the marine ecosystem, these ostracods have been reported to be incredibly useful for studying the physiology and biochemistry of bioluminescence.

Luminescent Courtship Displays of Three *Cypridinid ostracod* Species at Night.

Image credit: K.lousi. (Applied for permission)

Order: Myodocopida Family: Cypridinidae

Cypridina noctiluca

Common name (s): Not reported

Global distribution: It has global marine distribution

Habitat: Its occurrence is ranged from the intertidal to depths of 4500m

Biology: In the species of this genus, the fifth limb of female is without quadrate tooth.

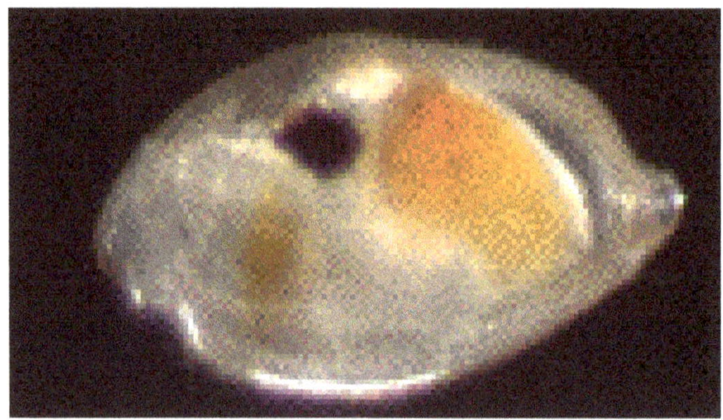

Image credit: Heterocycles. (Applied for permission)

An upper lip is divided in glandular, lobe-formed field, but it is without larger processes. Its 2nd antenna has a short, reduced, unjointed endopodite

Bioluminescence: Markova and Vysotski (2015) reported that bioluminescence of this species is due to the reaction between its *Cypridina* luciferin and *Cpyridina noctiluca* luciferase which is a secreted luminescent protein. According to Oba *et al.* (2007) this species is able to synthesize the *Cypridina* luciferin de novo from three amino acids *viz.* L-tryptophan, L-isoleucine, and L-arginine.

Enewton harveyi (= *Vargula harveyi*)

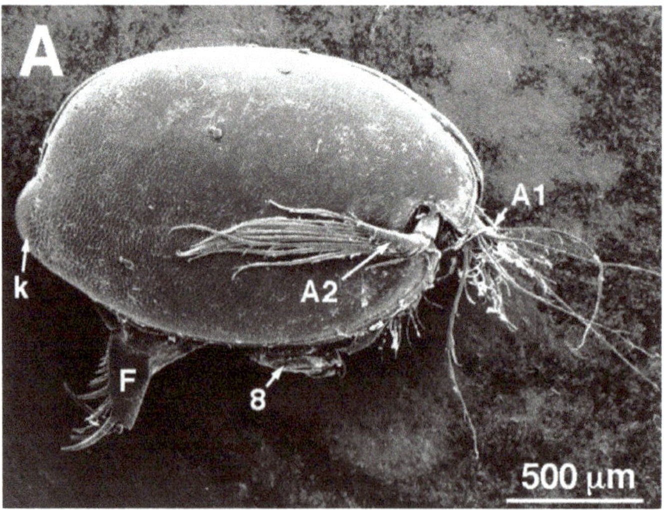

Image credit: James G. Morin, Reproduced with permission

Common name (s): Not reported

Global distribution: West Indian region; Jamaica, Puerto Rico, Bahamas

Habitat: It occurs in the seagrass beds nearby coral reefs at a depth range of 1.5-5m.

Biology: Valve shape of this species is round oval and male's keel shape is low pointy. Body of the animal is translucent without pigments in the valves or limbs. Gut is cream, tan, or reddish. The 16 ommatidia of its blackish lateral eyes are relatively small. While the length of the males varies from 1.4 to 1.5 mm, females from 1.7 to 2.0 mm.

Bioluminescence: The light organ of this species is comparatively small. It is very narrow proximally and wide distally. The displays of this species have been reported to occur as short, slow downward vertical-shortening displays without a trill low over grass beds. Each display begins only about 15-25 cm above the grass bed and it consists of a vertical-shortening train of 8 to 10 bright pulses which are often terminating at the bottom. These trains are about 15-25 cm in length; and are slow with a duration of about 15-20 s. (Cohen and Morin,2010)

Kornickeria louisi, Kornickeria marleyi, Kornickeria coufali* and *Kornickeria hastingsi

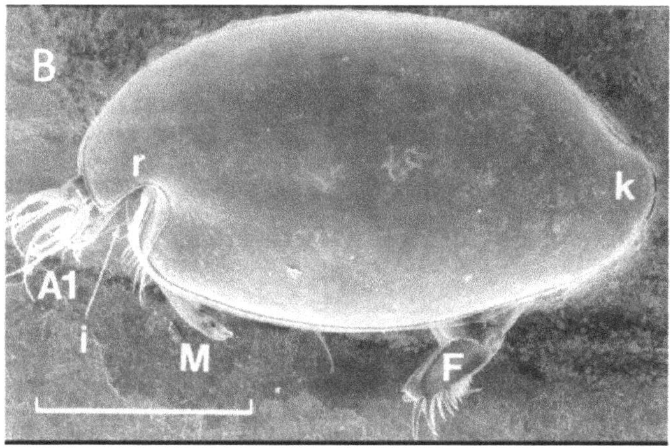

Kornickeria louisi

Image credit: James G. Morin, Reproduced with permission

Common name (s): Not reported as they are new species and new genus

Global distribution: All these species are found distributed in Caribbean Sea. While *Kornickeria louisi, Kornickeria marleyi,*and *Kornickeria coufali* are known only from Jamaica, *Kornickeria hastingsi* is from Jamaica, Belize and Roatan.

Habitat: These species occur in shallow coral reef habitats and sea grass beds at a depth range of 1-15m.

Biology: Valve shape of these species is elongate oval and the males' Keel shape is high rounded, with apparent inflexion. The presence of a paired male copulatory organs as an eighth pair of limbs in these species is a characteristic feature of this genus. While the length of the males of these species varies from 1.4 to 1.8 mm, females from 1.7 to 2.0 mm.

Bioluminescence: In all these species, the males have been reported to produce vertical trains of luminescent secretions as species-specific courtship patterns. Displays are known to be restricted to upward and downward variations of vertical displays and consist of a shortening phase of a few bright pulses and a longer trill phase of dimmer, fairly closely and evenly spaced pulses. Train lengths are also relatively short (Cohen and Morin (1993)

Maristella spp.

Maristella, a new ostracod genus in the family, Cypridinidae and it is the only genus in which all the males of its 7 species are known for their species-specific and lateral or diagonal luminescent courtship displays. These tiny animals are nicknamed as 'Star of the Sea' seed shrimp as they are twinkling like little stars in the ocean. In each species, males develop swirling trains of light in the water and pulse the same pattern. Amazingly the different species of this genus have distinct train patterns. However, surprisingly in all these species, females use their bioluminescence only to deter predators, but not in mating. It is believed that bioluminescence protects these ostracods by startling would-be predators with a blinding flash of light (https://cals.cornell.edu/news/cornell-discovered-shrimp-makes-2019-top-10-not-taste)

Maristella chicoi

Common name (s): "Star of the Sea" seed shrimp

Global distribution: Caribbean Sea

Habitat: It is found distributed everywhere in the ocean, from shallows to deep seas and in both tropical and polar waters.

Biology: Valve shape of these species is oval. Posterior D margin gently is sloped;. Males are with wide, somewhat rectangular keel which extends two-thirds height of the valve. Carapace and limbs of this species are transparent and without pigmentation. While the length of the males is varying from 1.6 to 2.2 mm, the females from 1.5 to 1.7 mm.

Bioluminescence: Its light organ is apparent and appears as a bright yellow rectangular at the anterior of the body. Male displays are believed to be oriented horizontal to the substrate and are showing high levels of entrainment. Further, the displays run in near-parallel bifurcations resulting in spectacular fan-like radiations of light pulse trains. It is reported that this species use its bioluminescence in courtship (Reda *et al.,* 2019; https://cals.cornell.edu/news/cornell-discovered-shrimp-makes-2019-top-10-not-taste)

Maristella ignitula, Maristella lucidella, Maristella micarnacula, Maristella noropsela, Maristella psammobia and *Maristella scintilla*

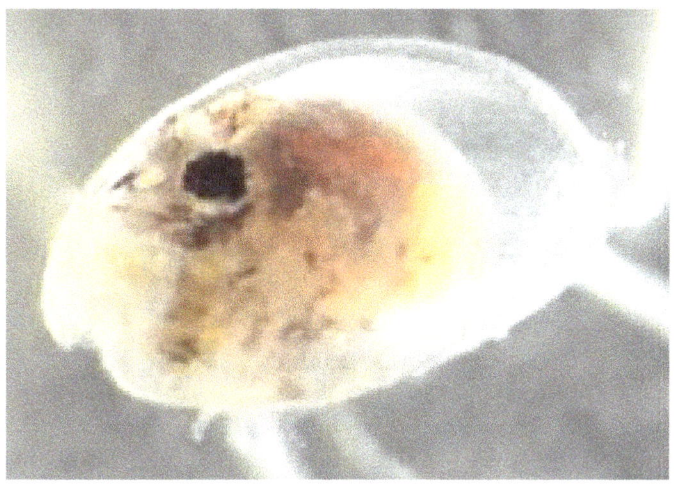

Maristella scintilla

Image credit: iNaturalist. (Applied for permission)

Common name (s): Not reported as all these species are new

Global distribution: All these species are found distributed in Caribbean Sea; San Bias Islands, Panama

Habitat: All these species have been collected from reef, sand near reef, and channels at a depth range of 0–30 m.

Biology: In all these species, valve shape is oval and the males' keel shape is intermediate rounded. While the length of the males is varying from 1.6 to 2.2 mm, the females from 1.5 to 1.7 mm.

Bioluminescence: Males of these species secrete pulses of luminescence in species-specific trains in the vicinity of shallow coral reefs at night. Cohen and Morin (1989) reported that these species differ in their bioluminescent signaling patterns as detailed below

Maristella psammobia (= *Vargula psammobia*) produces a series of pulses in the shape of an "L," low over sand patches near reefs.

Maristella ignitula (= *Vargula ignitula*) produces vertical downward displays over grass beds,

Maristella scintilla (= *Vargula scintilla*) produces vertical downward displays over slopes or patches of coral

Maristella micamacula (= *Vargula micamacula*) produces vertical downward displays at the edge of reef crests.

Maristella lucidella (= *Vargula lucidella*) produces vertical upward displays along reef crests.

Maristella noropsela (= *Vargula noropsela*) generates horizontal to long oblique trains over sand channels in the reef.

Photeros annecohenae (= *Vargula annecohenae*)

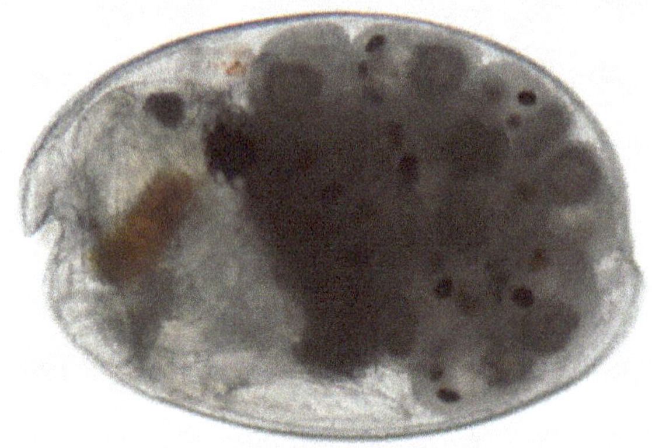

Image credit: Elizabeth Torres (CC)

Common name (s): Not known

Global distribution: Caribbean Sea off the coast of Belize

Habitat: It remains primarily on or in the shallow seabed

Biology: It is a small-bodied (<"2 mm) bioluminescent ostracod with an outer bivalve carapace composed of chitin and many pairs of differing appendages

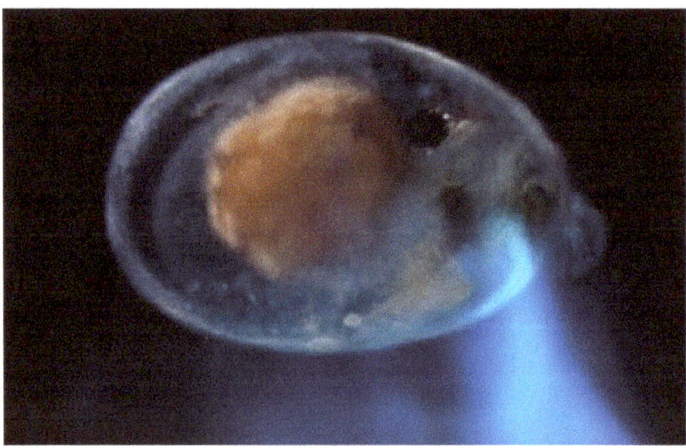

Adult Female *Photeros annecohenae* Secreting Bioluminescence.

Image credit: Elliot Lowndes, Reproduced with permission

Bioluminescence: This animal contains a light organ that creates chemicals that produce luminescence. Groups of males have been reported to synchronize their displays in a process called entrainment, creating captivating light shows of luminescence. When attacked these animals are known to release a cloud of luminescence that shocks and blinds its predator (Anon., http://bioweb.uwlax.edu/bio203/2011/kramolis_kali/). Adult males enter the water column nightly where they produce an elaborate species-specific display in order to attract females. These displays occur as a series of rapid flashes in an upward direction, above shallow grassbeds (Kates, 2011; Torres,2009; Gerrish and Morin,2008)

Photeros graminicola (= *Vargula graminicola*)

Common name (s): Not known

Global distribution: San Blas, Panama.

Habitat: It occurs in shallow sea grass beds at a depth range of 3-10 m.

Caribbean Panama; shallow sea grass beds

Biology: Valve shape of this species is rather boxy oval and high. Posterior dorsal margin of the valve is sloped, but it is more steeply than anterior dorsal margin, with low rather curved projecting keel. Average size of valve in male is 1.7 mm and 1.9 mm in female.

Bioluminescence: Cohan and Morin (2010) reported that its upward species-specific displays are composed of a series of clusters of rapid flashes and such displays are produced by swarms of vertically rising males above grass beds. Heger *et al.* (2007) reported that it produces two different kinds of luminescent displays. Firstly, the anti-predator release of a bomb-like cloud of luminescence, and secondly, complex trains of precisely spaced pulses emitted mainly by displaying males at lekking sites.

Photeros jamescasei

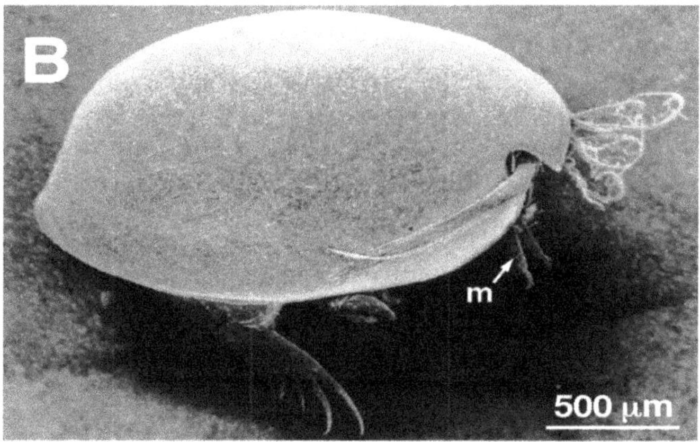

Image credit: James G. Morin, Reproduced with permission

Common name (s): Not reported

Global distribution: Jamaican coast

Habitat: This coral reef species has been collected only above the terraces of well-developed coral spurs at bottom depths of about 15-17 m

Biology: Valve shape of this species is rather boxy oval and high with Postero-dorsal curved corner. Males are with low rather pointed projecting keel. Valve length is varying from 1.9 to 2.1 mm in males and 2.4 mm in female. Color of the animal is very pale and transparent; valves and limbs are without pigment; and diatoms appear occasionally in the marsupial area. Gut is pale cream, tan, or light brown. Ommatidia of the eyes are relatively large.

Bioluminescence: In this species, its light organ is slightly curved, bar-shaped, and golden-yellow in coloration. It is known to give fairly long (1.5-2 m) and slow (15-20 s) upward vertical-shortening displays over coral reefs. Each display starts 1 to 2 m above the bottom and it is consisting of 6 -10, bright pulses followed by about 5-10 closely-spaced dimmer pulses. Pulse durations are relatively long and are ranging from 6 to 10 s (Cohen and Morin,2010)

Photeros johnbucki

Common name (s): Not reported

Global distribution: It is found distributed in Caribbean Sea; near Discovery Bay on the north coast of Jamaica

Habitat: These species occur in fore reefs and sea grass beds at a depth range of 1-30m.

Biology: In females, valve shape is rather boxy oval and high with postero-dorsal curved sloped corner. Males' keel shape is low pointy. Average valve length is 2.3 mm. Color of the animal is pale and transparent. Valves and limbs are without pigment. Gut is yellowish.

Bioluminescence: Its light organ is slightly curved, bar-shaped, and yellow with a slight greenish tinge. It makes its displays only over coral and sand near the border between large and flat sand areas. Its displays are known to occur as wide oblique downward arcs of 1-2 m length. Each display starts about 2- 3 m above the bottom and it consists of a train of 15-20 bright pulses. Train length is about 0.8 to 1.8 m long and it lasts about 15-25 s (Cohen and Morin, 2010)

Photeros mcelroyi

Common name (s): Not reported

Global distribution: Discovery Bay, North coast of Jamaica.

Habitat: It occurs on the deeper fore reefs at depths about 10-18 m

Biology: Valve shape of the male animal is oval. Posterior dorsal margin is sloped, as much or more than anterior dorsal margin. Males possess low and pointed projecting keel which is extending almost to mid height of valve. Animals are pale in coloration and the valves and limbs are without pigment. Gut is yellowish-white to cream and lateral eyes are maroon-brown. Length of male animals varies from 1.8 to 1.9 mm

Bioluminescence: The light organ of this species is bar-shaped, somewhat curved, and golden-yellow. Its displays are often observed in an upward direction above the tops and flat terraces of coral spurs. It is also reported that its displays form a series of rapid, strobe-like flashing signals that initially appear about 1 to 2 m above the bottom, followed by a few flashes occurring progressively higher in the water column. Further, each display consists of about 15-20 flashes, and the train duration ranges from about 15 s to almost 30 s,(Cohen and Morin,2010; Kates,2011)

Photeros morini (= *Vargula morini*)

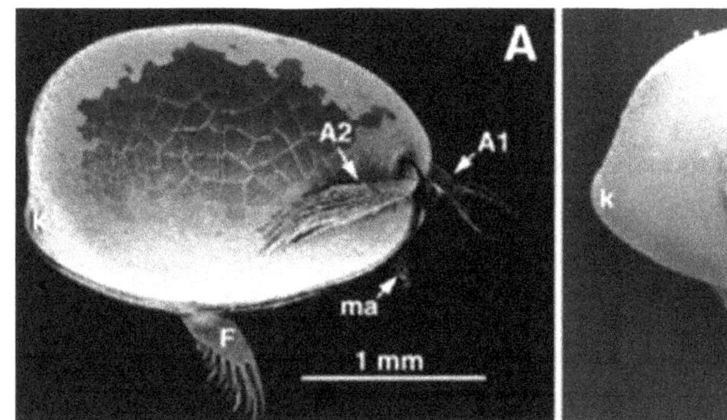

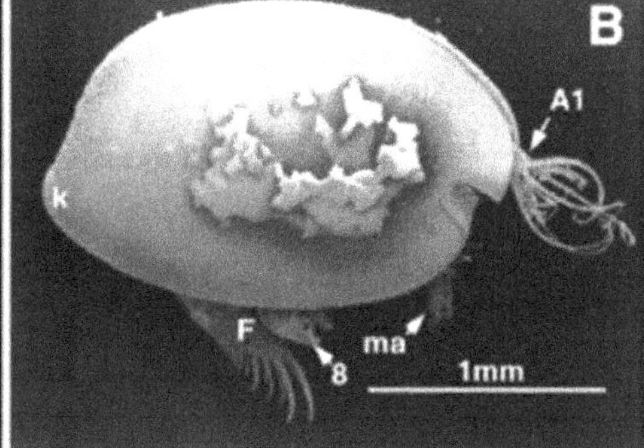

A: Female; B: Male

Image credit: Elizabeth Torres, Anne C. Cohen (Applied for permission)

Common name (s): It is a new species

Global distribution: It is commonly found distributed in the Carrie Bow Cay, Belize (a Caribbean country on the northeastern coast of Central America)

Habitat: It occurs in reef habitats including shallow fore-reefs at a depth of about 3–4 m; and back reefs with considerable water flow and depth of about 5–6m.

Biology: It is fairly a large species and the carapace length is varying from 1.88 to 2.02 mm in males; and females from 2.28 to 2.32 mm. Eyes are maroon-black in coloration and gut is normally army-green to light brown. A nicothoid copepod has been reported to live within the posterodorsal area of the valves of some individuals of this species.

Bioluminescence: It has a curved rectangular light organ which is light yellow in coloration. Its species-specific bioluminescent courtship display is a distinctive very rapid, strobe-like downward display which is usually consisting of about

10–20 flashes per train. Males' displays often occur over low mixed coral mounds of *Montastrea annularis, Siderastrea rubble* and over *Acropora cervicornis* thickets (Torres and Cohen,2005).

Photeros parasitica

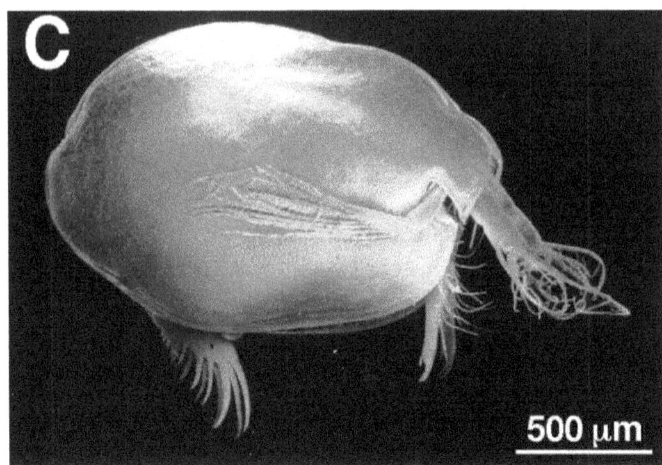

Image credit: James G. Morin, Reproduced with permission

Common name (s): Not known

Global distribution: North coast of Jamaica

Habitat: It occurs in shallow grass beds at depths of 1.5-5 m.

Biology: Valve shape of this species is oval and its postero-dorsal margin is sloped more than antero-dorsal margin. Males possess slightly projecting crescent-shaped keel which is extending slightly above mid-height of valve; and females are with a keel which projecting very little. Valves and limbs of this species are without pigment and gut is tan to white.

Bioluminescence: The light organ of this species is golden yellow and bar-shaped with a slight curve. Though the displays of this species have not been documented, it is known to produce a bright amorphous anti-predatory cloud of bioluminescence (Cohen and Morin,2010)

Photeros shulmanae (= *Vargula shulmanae*)

Common name (s): Not reported

Global distribution: San Blas, Caribbean Panama.

Habitat: It occurs in shallow seagrass habitats

Biology: Its valve shape is rather boxy oval and is high with postero-dorsal curved corner. Males possess low and pointed projecting keel. While the carapace length of the males is from 1.8 to 2.0 mm, that of females is from 2.3 to 2.6 mm.

Bioluminescence: Its displays are known to occur among large gorgonians on steep reef slopes. At night, male ostracods of this species have been reported to produce an elaborate species-specific display in order to attract females. These displays are known to occur as a series of rapid flashes in an upward direction on reef walls.(https://cypridinidae.myspecies.info/taxonomy/term/74/descriptions).Heger *et al.* (2007) reported that it produces two different kinds of luminescent displays. Firstly, the anti-predator release of a bomb-like cloud of luminescence, and secondly, complex trains of precisely spaced pulses emitted mainly by displaying males at lekking sites.

Vargula contragula

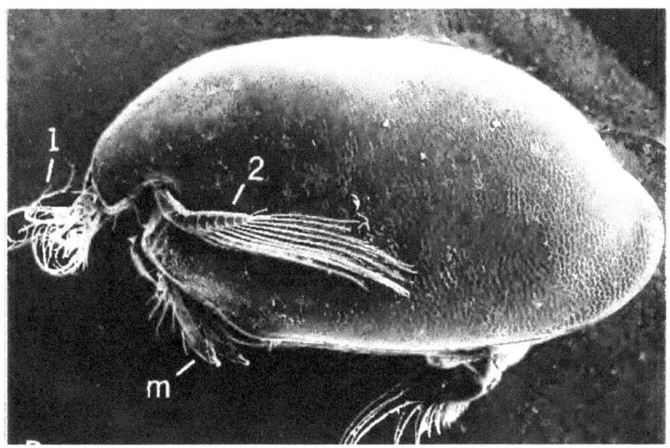

Mikroconchoecia echinulata

Image credit: World Ostracoda Database (CC/NC)

Common name (s): Not reported

Global distribution: West Caribbean; San Bias Islands

Habitat: It occurs in coral reef walls and slopes at a depth range of 1-10m

Biology: Valve of this species is elongate oval with a projecting bulge (bump) at its anteroventral corner. In males, the keel shape is high rounded. Eye is dark brown or maroon with 16 ommatidia. Colour of the animal is transparent and abdomen is orangish. Male's vale length is from 1.7 to 2 mm.

Bioluminescence: Its light organ is orange-brown. The display of males has been reported to occur 1-3 m over gorgonians in an oblique upward direction about 50 min after sunset. Pulse duration is 4 sec and is blue in color; pulse distance is 15 cm between first two pulses, 8 cm between 2nd and 3rd pulses, 5 cm between 3rd and 4th pulses, and 2 cm for the remaining pulses (Cohen and Morin 1986). Heger *et al.* (2007) reported that it produces two different kinds of luminescent displays. Firstly, the anti-predator release of a bomb-like cloud of luminescence, and secondly, complex trains of precisely spaced pulses emitted mainly by displaying males at lekking sites.

Vargula hilgendorfii

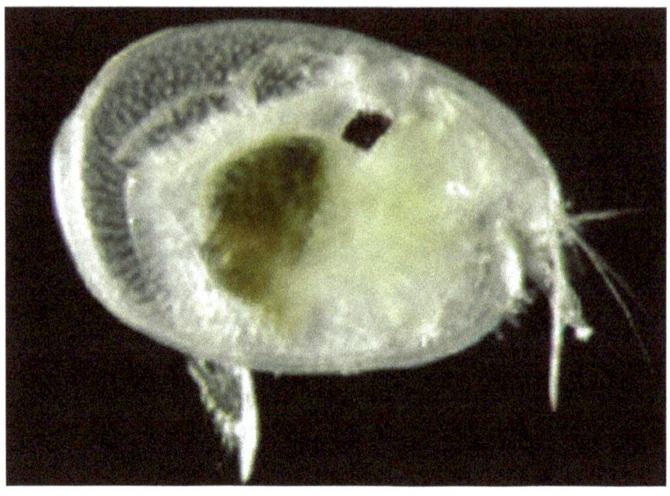

Male

Image credit: Yoshi, Reproduced with permission

Common name (s): Sea-firefly

Global distribution: It can only be found in Japan, the Gulf of Mexico, the Californian Coast and the Caribbean Sea.

Habitat: It lives in the sand at the bottom of shallow water. It is a nocturnal species.

Biology: These individuals are tiny, flat, seed-like creatures They range in size from 1 to 3mm.

Bioluminescence: It has a yellow luminescent organ in its upper lip region. It is known for the emission of its blue-coloured light by a specialized chemical reaction of the substrate luciferin (vargulin) and the enzyme luciferase. Its luciferase enzyme consists of a 555-amino acid-long peptide with a molecular mass of 61627 u, while the luciferine has only a mass of 405.5 u. The maximum in the wavelength of its luminescence is dependent on pH and salinity of the water in which the above reaction takes place. The wavelength has been reported to vary between 448 and 463 nm and the maximum 452 nm is with the sea water. In this species, the substrate vargulin ejected from the upper-lip gland oxidizes with luciferase as a catalyst. This reaction produces carbon dioxide, oxyluciferin, and the resulting blue light. (https://en.wikipedia.org/wiki/Vargula_hilgendorfii) These animals use their light displays for defense and in courtship.

Markova and Vysotski (2015) reported that bioluminescence of this species is due to the reaction between its *Cypridina* luciferin and *Cpyridina* luciferase (type) and the wavelength of emission maximum was found to be 465 nm. According to Oba *et al*. (2007) this species is able to synthesize the *Cypridina* luciferin de novo from three amino acids *viz.* L-tryptophan, L-isoleucine, and L-arginine.

Sea-Firefly Glowing

Image credit: Almandine (Wikimedia Commons)

These animals when crushed in sand and water,they emitted blue luminescence which was used by soldiers in World War II to provide a dim night-light (https://www.carolina.com/cellular-physiology-enzymes/sea-firefly-cypridina-hilgendorfii/FAM_203430.pr).

In Japan, these animals were dried and ground into a bio-luminescent powder that provided portable light sources during wartime. In the Caribbean, these animals are called as 'blue tears' due to their light streams left in the night ocean. The appearance of the swarming of this species has been reported as a source of wonder and delight in the summer nights.(https://chapinthehat.com/sff-video)

Vargula kuna and *Vargula mizonomma*

Common name (s): Not reported

Global distribution: Southwest Caribbean

Habitat: They occur in coral reef areas at a depth range of 0-12m.

Biology: The valve shape of these species is oval with a length varying from 1.7 to 1.8 mm in males and females ha e an average length of 1.7 mm. Keel shape of the males of these species is intermediate rounded.

Bioluminescence: Males of these ostracod species have been reported to secrete pulses of luminescence in species-specific trains above shallow coral reefs at night. While the pattern of display in *Vargula kuna* is vertical as either long upward or shorter downward trains, the pattern of *Vargula mizonomma* is zigzag and upwards (Morin and Cohen,1988)

Vargula norvegica

Common name (s): Not known

Global distribution: NE Atlantic Ocean and it is widespread along the west coast of Norway

Habitat: This benthic species occurs in areas of mud and sandy mud substrata at a depth range of 73–680 m

Biology: Valve shape of this species is round oval and male's keel shape is low, short crescent. While the length of the males vary from 3.3 to 3.7 mm, females from 3.3 to 3.8 mm.

Bioluminescence: This species luminesced readily in response to mechanical stimulation and it was also observed emitting sufficient volume of luminous secretion. Its luminescence has been reported as a defence response against the predatory activity of eel *Synaphobranchus kaupii* which is believed to compete for its bait and also feed on this species (Heger *et al.*, 2007)

Vargula tsujii

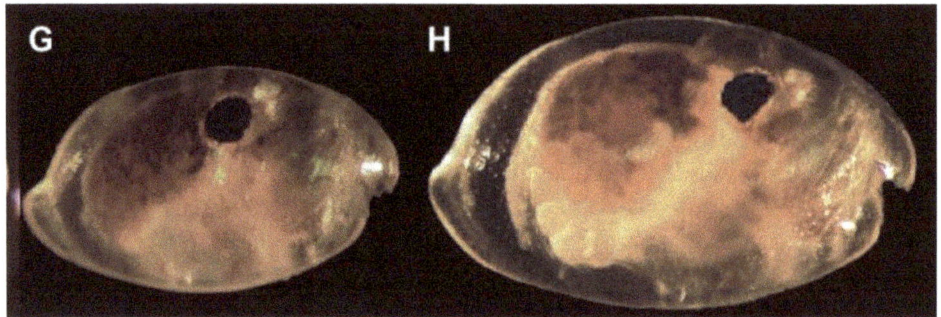

G: Male H; Female

Image credit: Todd Oakley, Reproduced with permission

Common name (s): Not reported

Global distribution: Pacific coast of North America; Baja California to Monterey Bay

Habitat: This species is demersal and it is common in sandy areas at a depth of 3–931 m. It has been reported to move between the benthic and planktonic environments and is primarily active in or near the kelp forest at night

Biology: Its valve is elongate oval and male's keel shape is low, rounded, projecting bump. While the length of males is varying from 1.7 to 2.3 mm, females range from 2 to 2.3 mm. It has a short life cycle and is remarkably bright compared to other bioluminescent organisms.

Bioluminescence: Unlike other bioluminescent organisms, this species has been reported to synthesize its own luciferins (https://dailynexus.com/2020-07-23/bright-ideas-ucsb-researchers-establish-a-laboratory-culture-of-bioluminescent-ostracods/).This species serves as an important prey item of the plainfin midshipman fish (*Porichthys notatus*), as this ostracod is the source of luciferin for the bioluminescence seen in the fish (Goodheart *et al.*, 2020)

Family: Halocyprididae

Mikroconchoecia echinulata

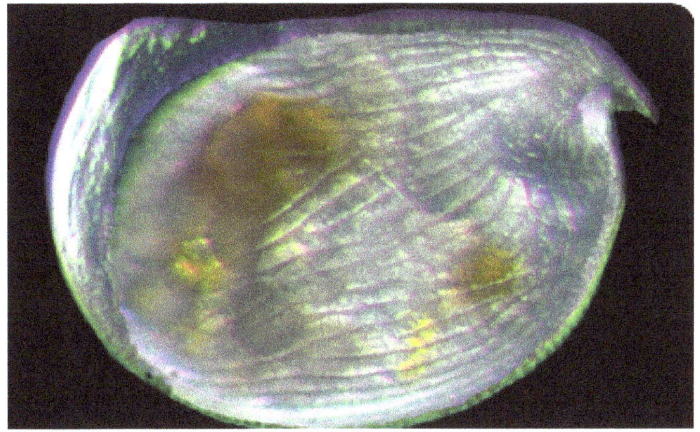

Image credit: Hopcroft, Russell WoRMS, Reproduced with permission

Common name (s): Not known

Global distribution: Atlantic and Indian Oceans

Habitat: It is a shallow epi- mesopelagic species

Biology: It is distinguished from its related species by its more arcuate posterior margin of the carapace. Carapace length of male and female is varying from 0.78-0.8 mm and 0.75-0.8 mm respectively

Bioluminescence: In this species bioluminescence glands are present on the lower edge of the incisure and also where the posterior margin is arched into an exhalent siphon. These bioluminescence glands have been reported to release bioluminescence into the respiratory flow of water. This flow is known to enter the carapace below the rostrum, and is discharged through the exhalent siphon on the posterior margin (https://www.nhm.ac.uk/research-curation/scientific-resources/biodiversity/global-biodiversity/atlantic-ostracods/ostracods/structure/glands.html)

Conchoecilla daphnoides

Common name (s): Not known

Global distribution: It is found distributed in all oceans.

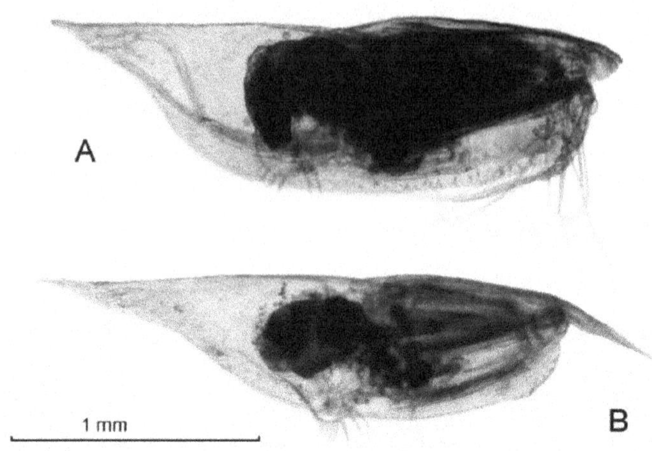

A: Female; B: Male

Image credit: WoRMS. (CC/NC)

Habitat: It is an epipelagic to deep mesopelagic species and it is known for its vertical migration.

Biology: In this species, its postero-dorsal region of the carapace is developed into a long spinose pointed process. In the female, its carapace length is far more elongated (about 5 times more than the height) than in male. Further, it has a long and pointed rostrum, which is longer on the left valve. On the other hand, the carapace of male is elongated with serrate posterior margins on both valves. Carapace length of the female and male is varying from 4.2 to 5.9 mm and from 2.3 to 3.3 mm respectively

Bioluminescence: In the presence of bioluminescence glands, this species resembles *Mikroconchoecia echinulata*. Apart from these glands, additional bioluminescence -generating glands have been reported to be present in the tip of the spine of its elongated rostrum (https://www.nhm.ac.uk/research-curation/scientific-resources/biodiversity/global-biodiversity/atlantic-ostracods/ostracods/structure/glands.html)

Conchoecissa symmetrica

Common name (s): Not known

Global distribution: It is largely confined to the Southern Ocean; North Atlantic

Habitat: it is a deep mesopelagic to bathypelagic species usually occurring at depths >750m.

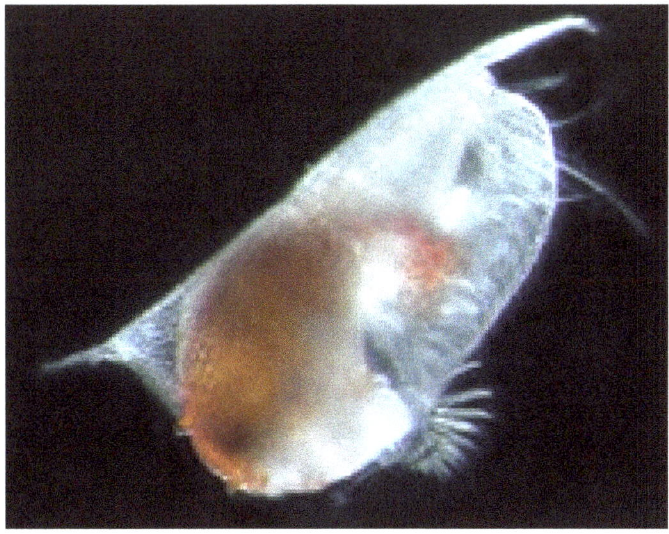

Conchoecissa sp.

Image credit: Copepodia. (Applied for permission)

Biology: In this species, its carapace is highest posteriorly and surface of carapace is strongly reticulated. Further its rostrum is pointed, long and bent downward. It grows to a maximum length of only 4 mm.

Bioluminescence: In the presence of bioluminescence glands, this species resembles *Mikroconchoecia echinulata.* Apart from these glands, additional bioluminescence -generating glands have been reported to be present at the tips of large spines or tubercles at the posterior or ventral dorsal corners of the rostrum (ttps://www.nhm.ac.uk/research-curation/scientific-resources/biodiversity/global-biodiversity/atlantic-ostracods/ostracods/structure/glands.html).

3.6.4. Decapod Shrimp

Subphylum: Crustacea; Class: Malacostraca; Order :Decapoda

The order Decapoda has more than 10,000 species. This order has several infraorders *viz.* Penaeidea (marine shrimp), Caridea (marine, estuarine and freshwater shrimp), Astacidea (lobsters, crayfish), Palinura (spiny lobsters), Anomura (mud-borrowing shrimp, hermit crabs, other) and Brachyura (marine, estuarine and freshwater crabs).Among these decapods, the pelagic decapod shrimps have been reported to play a significant role as food for other shrimps, squids, fish and marine mammals. Besides playing a significant role, these decapod shrimps serve as a link between mesoplankton and higher trophic levels; and as agents in the movement of organic matter to the deep sea through their vertical migrations. Among these decapod shrimps, the oplophorids, including the genera *Systellaspis, Acanthephyra,* and *Oplophorus,* are known to disgorge large volumes of luminous fluid. The visual systems of these crustaceans have been recently investigated in relation to their bioluminescent ability and vertical migrations (Haddock *et al.,* 2010)

Family: Acanthephyridae

Acanthephyra brevirostris

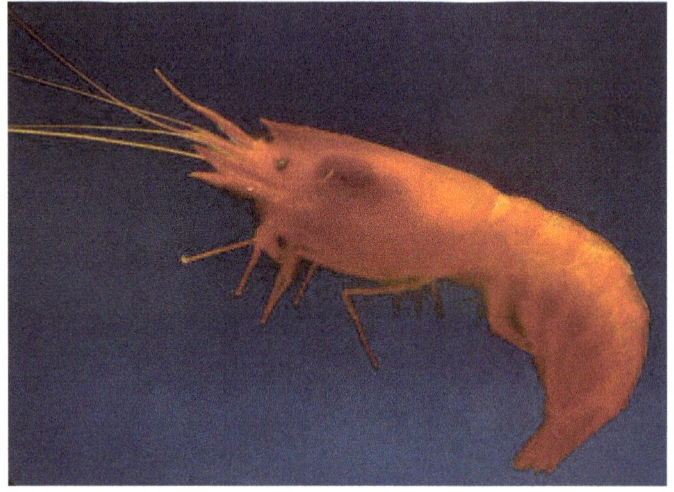

***Acanthephyra* sp.**

Image credit: NOAA (Public Domain)

Common name (s): Carid shrimp

Global distribution: A near-cosmopolitan species of tropical and temperate waters; Gulf of California, Mexico to Ecuador (eastern Pacific); Atlantic Ocean, SW Indian Ocean

Habitat: This bathypelagic species has a depth range of 1200 - 5394 m

Biology: This bilaterally symmetrical, fragile species has a soft carapace and its rostral teeth are saw-like. posteromedian tooth on 3rd pleonic somite is fleshy and is overreaching 4th somite.

Bioluminescence: This species has been reported to emit light from its liver extracts and the emission maxima have been observed as 445-460 nm (Herring (1981)

Acanthephyra curtirostris

Common name (s): Not known

Global distribution: Southwestern Atlantic

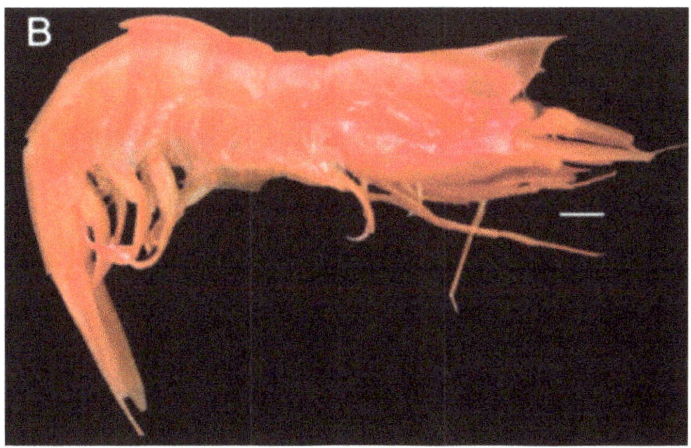

Image credit: Flavio de Almeida et al. (Applied for permission)

Habitat: This deep-sea shrimp is commonly found in seamounts at the depths of 50 -1200 m during day and 500-1200 m during night

Biology: Carapace of this species is not dorsally carinate posteriorly. Abdomen is carinate on all but the first somite. Telson is dorsally sulcate on proximal half and is armed with 6–15 dorsolateral spines. Its integument is firm and rostrum is not reaching beyond antennular peduncle.

Bioluminescence: Its spew and liver extracts are bioluminescent.The emission maxima of this species have been reported as 455-460nm (Herring, 1981). Frank and Case, (1988) observed that the maximum spectral sensitivity of this species was at 510 nm, a longer wavelength than expected for such deep-sea dwellers.

Acanthephyra eximia

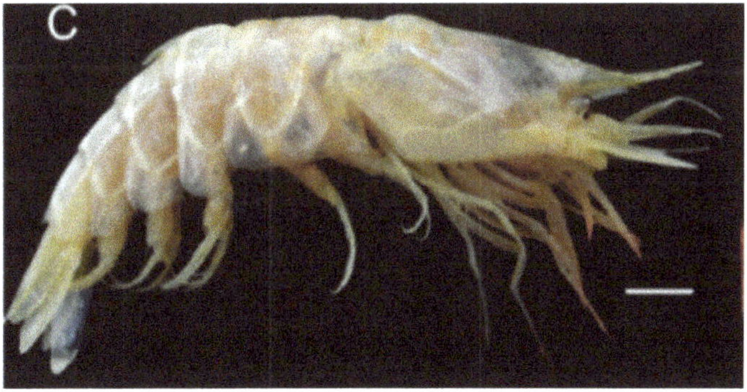

Image credit: Flavio de Almeida et al. (Applied for permission)

Common name (s): Not reported

Global distribution: Southwestern Atlantic; Indo-West Pacific from East Africa to Japan, Hawaii and Australia (WA)

Habitat: This bottom dwelling species is commonly found in seamounts at the depths of 50–4700 m

Biology: In this species, its carapace with rostrum is overreaching scaphocerite. Ventral margin is with 3–4 teeth. Antennal spine is present. Branchiostegal spine is present without distinct carina. Abdomen is dorsally carinate on all somite except somite 1. Male pleopod 1 is with endopod which is rounded with numerous stout setae on proximal to mesial portion. Female's carapace length is 36 mm

Bioluminescence: Its liver extracts possess luminescence. Herring (1981) reported that its emission maximum was at 450 nm

Acanthephyra pelagica

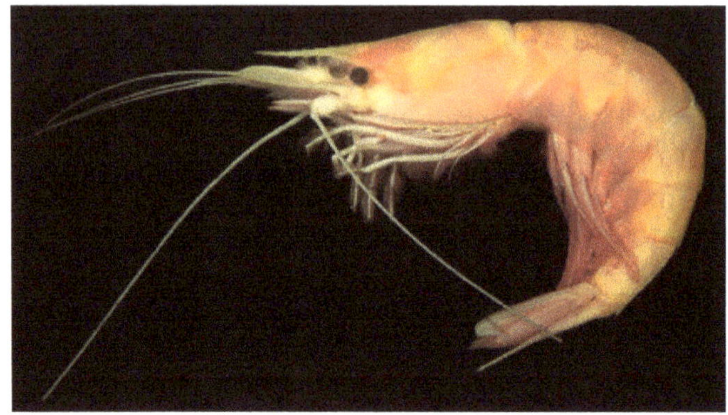

Image credit: Eric A. Lazo-Wasem - Gall L, Wikimedia Commons

Common name (s): Not reported

Global distribution: This cosmopolitan species occurs in the Atlantic, Pacific and Indian Oceans. It is common in boreal and subtropical zones of the North Atlantic.

Habitat: This bathy-mesopelagic species has a depth range of 183 - 2500 m. But it is commonly seen at the 700–1800 m

Biology: The presence of a telson with 7–11 pairs of dorsolateral spines is the characteristic feature of this species.

Bioluminescence: When handled, these animals discharged a copious luminous secretion which is believed to be from its liver extracts. The blue light so generated persisted for 3-4 sec in some individuals. However, on stimulation, its luminescence became weak at every discharge (Clarke *et al.*, 1962)

Acanthephyra purpurea

Common name (s): Fire -breathing shrimp, glowing shrimp or purple deep-sea shrimp

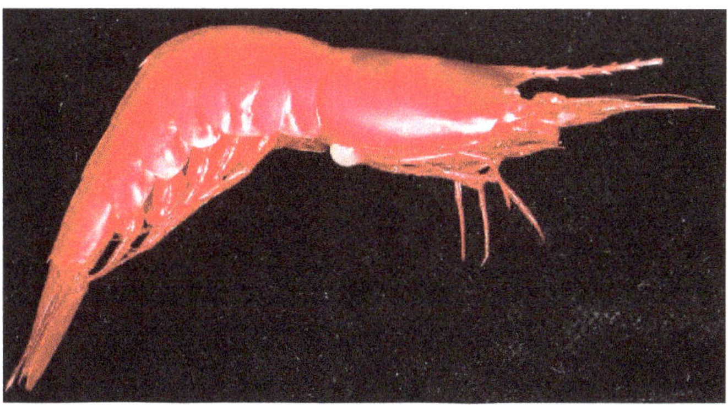

Image credit: INPN Paris with J. Poupin authorization, Reproduced with permission

Global distribution: North Atlantic, Gulf of Mexico, Caribbean Sea, Suriname and French Guiana; and tropical to polar (Arctic) climates of northern hemisphere.

Habitat: This pelagic species exists within the black void of the deep ocean at a depth range of 300 - 3292 m.

Biology: These shrimps are bright red in color owing to a heavy deposition of carotenoids in the body coverings. Eyes are black. Entire surface of the body is covered with greatly expanded chromatophores. Carapace is not dorsally keeled on posterior half. Rostrum is prominent and is toothed dorsally and ventrally. Branchiostegal spine is supported by only a very short keel. Cornea is wider than stalk. Second abdominal segment is keeled dorsally. Telson has 4–19 pairs of dorsolateral spines

Bioluminescence: When excited mechanically, this little shrimp has been reported to spew from its mouth bright blue bioluminescent cloud (fluid) which is originated in its hepatopancreas. When this fluid is discharged into the water, luciferin present in the fluid reacts with oxygen to produce luciferase and light. Herring (1981) and Nicol (1958) reported that its emission maximum was at 458nm and the luminescence exhausted under repeated stimulation. Its luminous fluid gets luminescent even if it is macerated (https://www.wired.com/2012/01/glow-little-spewing-shrimp-glow/). It is believed that its bioluminescent fluid is intended to mainly confuse or distract predators.

Acanthephyra sibogae

Common name (s): Not reported

Global distribution: Not available

Habitat: Not available

Biology: In this species, there is no carina on dorsal midline of its 2nd pleonic somite. Posteromedian tooth on 3rd pleonic somite is not fleshy, and is not overreaching 4th somite.

Bioluminescence: The luminescence of is species is due to its liver extracts and the emission maximum was 458nm (Herring, 1981)

Acanthephyra smithi

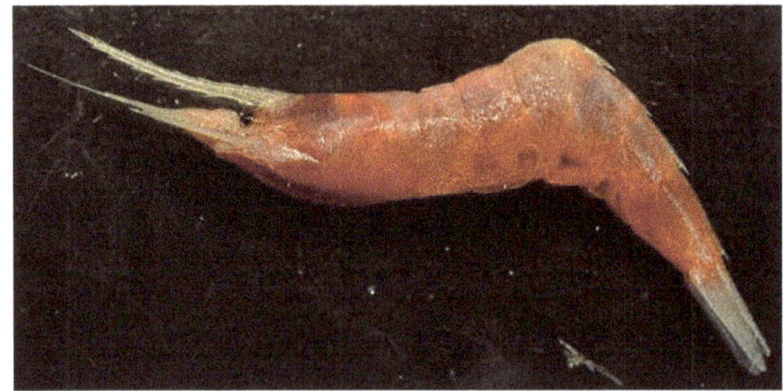

Image credit: Jeff Dubose. (Applied for permission)

Common name (s): Not reported

Global distribution: Indo-Pacific.

Habitat: This bathypelagic species has a depth range of 200 - 3716 m

Biology: Individuals of this species have a large and stout body which is covered uniformly with fine pits. Rostrum is usually shorter than the carapace. Branchiostegal spine is small and is supported by a very small carina. A distinct dorsal carina is present on the second to sixth abdominal somites. Carinae of the third to sixth somites ends in large teeth. Telson bears 3 pairs of dorsolateral spines. It has a maximum length of 2.5 cm.

Bioluminescence: Its spew is bioluminescent and this luminescent fluid is secreted from its hepatopancreas. Frank and Case, (1988) observed that the maximum spectral sensitivity of this species was at 510 nm a longer wavelength than expected for such deep-sea dwellers. Herring, (1981) and Poupin *et al.* (1999) however reported that its emission maxima were 455-460nm and 445-460 nm respectively

Acanthephyra stylorostratis

Common name (s): Not reported

Global distribution: Atlantic Ocean, Indo-Pacific, Antarctic and the Arctic: French Polynesia to USA and Canada.

Habitat: This bathypelagic species has a depth range of 700 - 5000 m

Biology: Carapace of this species has a thin crest that projects in a rounded rostrum from 4–6 spiniform teeth dorsally. Antennal spine is small. Its branchiostegal spine is supported by a long carina which is reaching 2/3 of carapace. Abdomen is dorsally carinate on all somites, except on somite 1. Somites 3 to 6 are with posteromesial tooth.

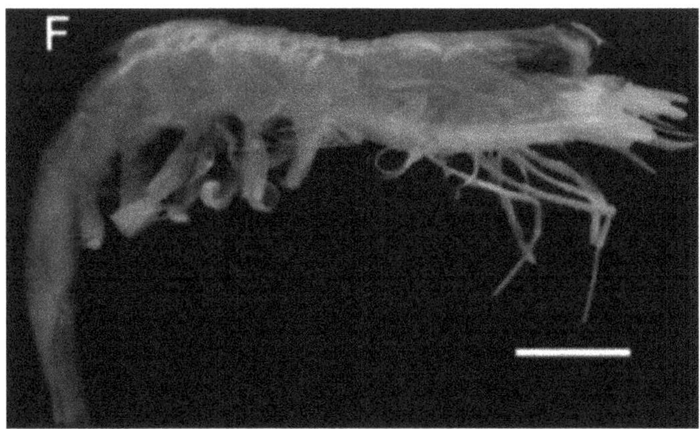

Image credit: Flavio de Almeida et al. (Applied for permission)

Bioluminescence: It largely resembles other species of this genus including *Acanthephyra smithi* in its bioluminescence. Herring, (1981) and Poupin *et al.* (1999) however reported that its emission maxima were 455nm and 445-460 nm respectively

Acanthephyra tenuipes (= *Acanthephyra gracilipes*)

Common name (s): Not reported

Global distribution: Southwestern Atlantic

Habitat: It is commonly found in seamounts between the depths of 50–1260 m

Biology: In this species, there is no carina on dorsal midline of 2nd pleonic somite. It s posterodorsal tooth on 3rd somite is low and is offset to left

Bioluminescence: It largely resembles other species of this genus as its bioluminescence is due to its liver extracts. Herring, (1981) reported that its emission maximum was 460nm.

Notostomus elegans

Common name (s): Not reported

Global distribution: Southwestern Atlantic

Habitat: This deep- sea decapod shrimp is commonly found in seamounts between the depths of 50–1260 m

Biology: In this species, its carapace with rostrum is overreaching scaphocerite length. ntennal spine and branchiostegal spine are present. Abdomen is dorsally carinate on all somites. Somites 3–6 have posteromesial tooth; and third somite is distinctly strong with dorsal carina. Female 's first pleopod is with leaf shaped endopod which has numerous plumose, articulated setae on lateral margin

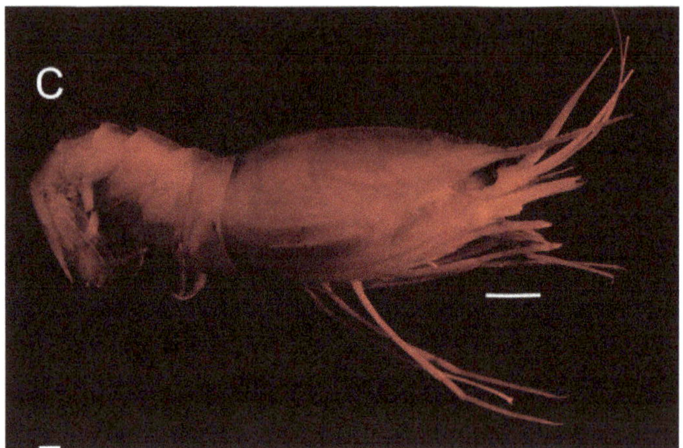

Image credit: Flavio de Almeida et al. (Applied for permission)

Bioluminescence: Its spew is bioluminescent and the luminescent fluid is secreted from its hepatopancreas. Frank and Case, (1988) observed that the maximum spectral sensitivity of this species was at 490 nm; and chromatic adaptation experiments indicated the presence of a single visual pigment.

Notostomus gibbosus

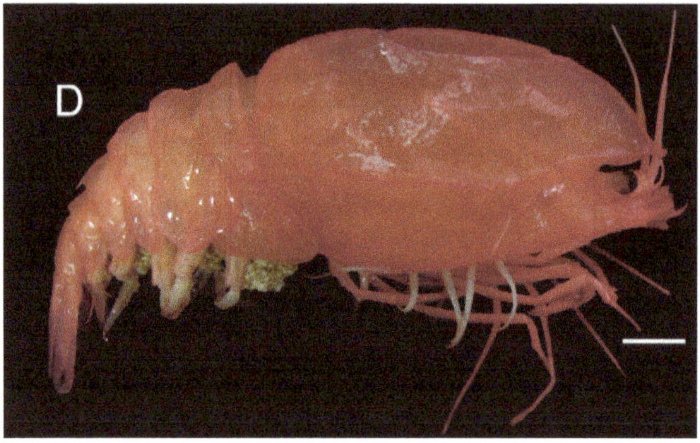

Image credit: Flavio de Almeida et al. (Applied for permission)

Common name (s): Not reported

Global distribution: Pacific, Atlantic and Indian Oceans

Habitat: This deep-sea shrimp is commonly found in seamounts between the depths of 50–1260 m

Biology: This micronektonic species is positively buoyant, possessing a dorsally enlarged carapace which contains a low-density fluid. Rostrum is with

and dorsal margin of its carapace is rather regularly convex in adults. Abdomen is with median dorsal carina on first somite

Bioluminescence: Its spew is bioluminescent due to the secretion of luminous fluid in its hepatopancreas. Frank and Case, (1988) observed that the maximum spectral sensitivity of this species was at 490 nm; and chromatic adaptation experiments indicated the presence of a single visual pigment.

Ephyrina figueirai

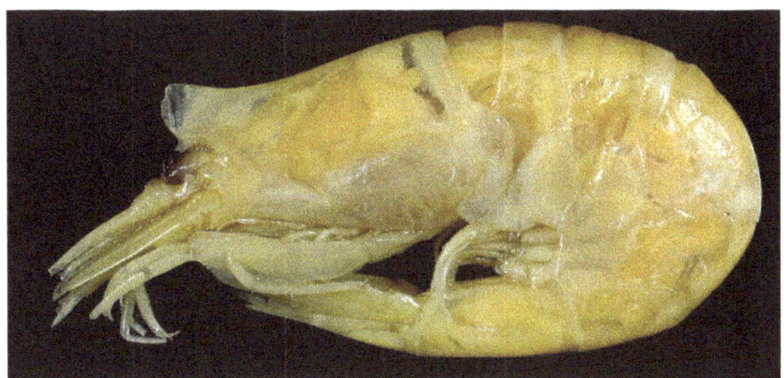

Image credit: Laura FLAMME - MNHN, Wikimedia Commons

Common name (s): Contour deep-sea shrimp, carid shrimp

Global distribution: Western Atlantic: USA and Canada; The Arctic.

Habitat: It is a deep-sea pelagic species

Biology: In this species, body is bilaterally symmetrical. Rostrum is laterally compressed into anteriorly truncate and dorsally unarmed crest. Its cornea is at least as wide as eyestalk and is darkly pigmented. 1st maxilliped with slender central lobe subdivided by 2 distinct transverse sutures. It has double rows of lateral spines on each side of the telson.

Bioluminescence: In this species, its light illuminate the eyes as the shrimp moves away backwards. In order to facilitate light emission, there is anterior/ posterior gradients of reflectivity and the presence of a hole in the posterior pat of the tapetum (Klein, 2000)

Heterogenys microphthalma (= *Acanthephyra microphthalma*)

Common name (s): Not reported

Global distribution: It has cosmopolitan in tropical and temperate seas;. It is commonly distributed in Bay of Bengal, southern Pacific and western North Pacific

Habitat: It is a bathypelagic to abyssopelagic species.

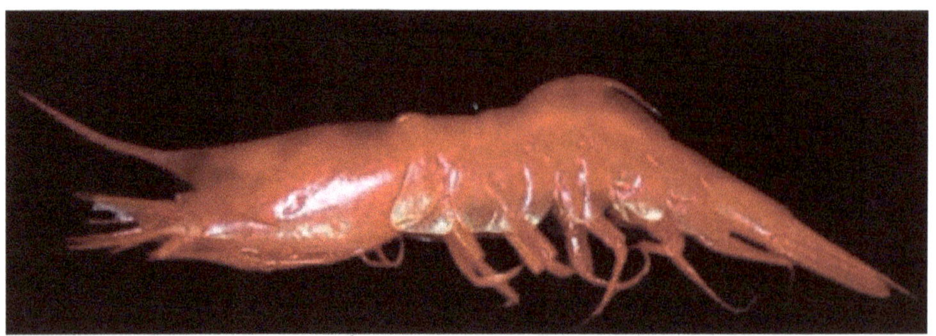

Image credit: Flavio de Almeida et al. (Applied for permission)

Biology: In this species, its rostrum bears fewer teeth dorsally than ventrally. Eyes have small cornea. Abdominal segments 3-6 are dorsally and 3rd segment has a slender, vry long median dorsal spine which reaches beyong the 4th abdominal segment.

Bioluminescence: As in certain species of this genus, its bioluminescence is largely due to its secretion of hepatopancreas. Herring, (1981) and Poupin *et al.* (1999) however reported that its emission maxima were 455,460 nm and 445-460 nm respectively

Family: Aristeidae

Cerataspis coruscans (= *Aristaeus coruscans; Plesiopenaeus coruscans*)

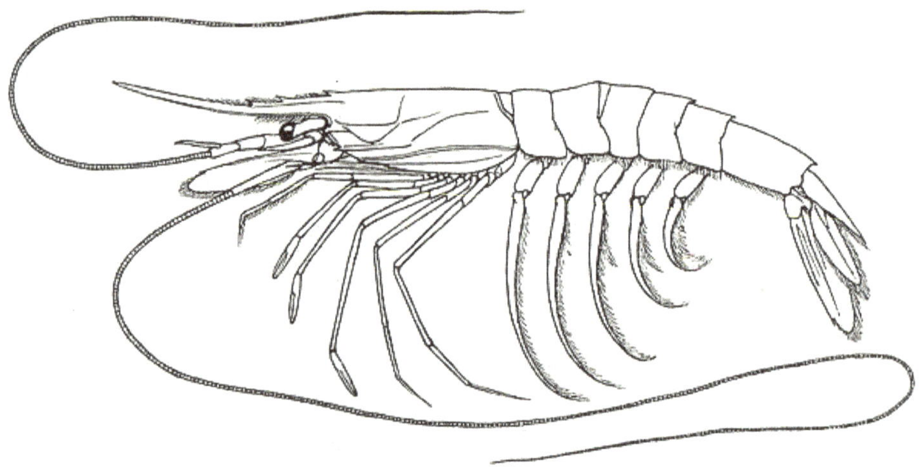

Image credit: Alfred William Alcock, Wikimedia commons

Common name (s): Not known

Global distribution: Indian Ocean: Northeast of Australia. Eastern Atlantic Ocean: Madagascar; Arabian Sea; Bay of Bengal; Andaman Islands.; Western Atlantic Ocean: Gulf of Mexico.

Habitat: This abyssopelagic species has a depth range of 900–2367 m.

Biology: In this brilliantly-luminous deep-sea prawn, rostrum is curved sligthly upward, with 3 dorsal spines,; and is with few simple setae on ventral region of rostrum and in front of rostrum spines

Bioluminescence: In this species, the luminous secretion is discharged from the glands at the base of the second pair of antennae (Nicol,1958).

Family: Euphausiidae

Nematobrachion flexipes

Common name (s): Not reported

Global distribution: Southwestern Atlantic, equatorial eastern-Pacific, California Current, and the Peru-Chile Current; eastern Gulf of Alaska; tropical and subtropical waters of the Indian Ocean

Habitat: It is commonly found in seamounts at depths of 50–1260 m

Biology: In this species, its carapace bears a small lateral denticle near the lower margin. Anterior part of the carapace has a well-developed keel. Rostrum extends to the anterior limit of the eye as a slender keeled extension of the frontal plate. Adults of this species have a length of 20-23 mm.

Bioluminescence: In this species, male has a prominent abdominal photophore which is believed to help in the emission of light (Schweiker *et al.*, 2020)

Family: Gnathophausiidae

Neognathophausia ingens (= *Gnathophausia ingens*)

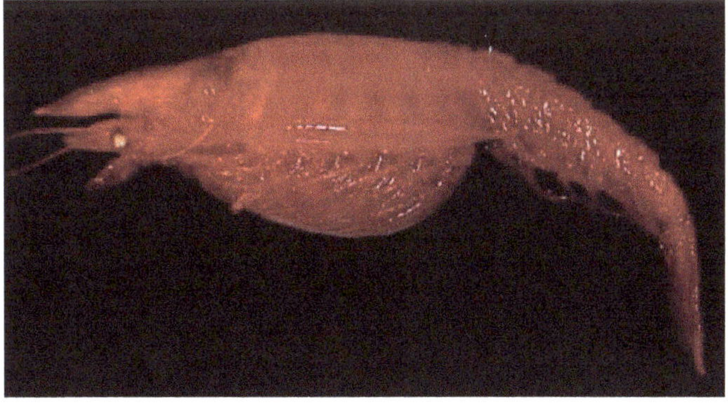

Image credit: CSIRO and photo credit to Karen Gowlett-Holmes; Katherine, Reproduced with permission

Common name (s): Deep water giant red mysid, or giant luminescent opossum shrimp

Global distribution: This cosmopolitan species has pantropical distribution. It is commonly distributed in the eastern Pacific Ocean off California, north of the Azores Islands, Gulf of Mexico and Bay of Bengal

Habitat: This bathypelagic species has a depth range of 175–1770 m

Biology: Body of this species is broad. Carapace is rigid. Rostrum is short, triangular and less denticulate in mature individuals. Posterodorsal spine is reduced and ventrolateral margins of the abdominal somites are produced into two spines. Telson is large, linguiform and is exceeding uropods. Apex of telson extends outwards forming lappets. It attains a maximum length of 210 mm.

Bioluminescence: This species has been reported to obtain its coelenterazine from its diet to help in the emission of light (Mirza and Oba,2021)

Family: Oplophoridae

Janicella spinicauda (= *Hoplophorus spinicauda*)

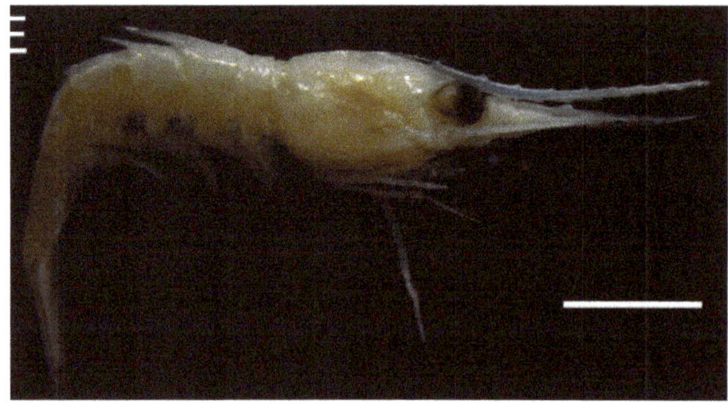

Image credit: Flavio de Almeida et al. (Applied for permission)

Common name (s): Not known

Global distribution: It has world wide distribution in tropical seas: Belize, Morocco, northern Madagascar, southern India, Hawaiian Islands, Florida, West Indies, Philippines, E. Indian Ocean and Japan.

Habitat: It is commonly found in seamounts at depths of 50–1260 m

Biology: In this species, body is slender and shell is flexible. Its carapace with rostrum overreaching scaphocerite; and dorsal margin is with 10–12 teeth and ventral margin has 6–7 teeth. Antennal spine is absent. Branchiostegal spine is not well marked and is without distinct carina. Rostrum is always with 8 teeth above and 7 teeth below. Abdomen is dorsally carinate on somites 2–4. Lateral surfaces of abdomen is with transparent, yellow-green, patches and tips of uropods are reddish. Telson is sulcate in dorsal midline and it has three pairs of dorsolateral stout setae.

Bioluminescence: In this species, the luminous secretion is discharged from its antennal glands or head glands (Nicol,1958). It is bioluminescent due to its spew and photophores which are present at various places on its body especially along the underside and sides. Its sensitivity maxima were at 400 and 500 nm, and chromatic adaptation experiments indicated the presence of two visual pigments (Frank ande Case, 1988). Herring (1981) reported that its cephalothorax emits a luminous substance and the emission maxima were at 456 and 458nm.

Oplophorus gracilirostris (= *Hoplophorus gracilorostris*)

Image credit: Flavio de Almeida et al. (Applied for permission)

Common name (s): Not known

Global distribution: It has world wide. Distribution and it is commonly distributed in Pacific and Indian Oceans and Japan

Habitat: It is commonly found in seamounts at depths of 50–2400 m. It is also found in underwater volcano.

Biology: Body of this species robust and shell hard. Though body is transparent it is covered with dense scarlet spots, especially in mid-lateral region of each abdominal somite and posterior margins of posterior three abdominal segments. Lateral surfaces of abdomen has transparent yellow-green patches. Eye is deep vermilion. Carapace with rostrum is overreaching scaphocerite and ventral margin is with six teeth. Antennal spine is present. Rostrum is extending beyond scaphocerite and is armed with 12–14 teeth dorsally and 9–11 teeth ventrally.

Bioluminescence: In this species, its biosynthesized and secreted Oplophorus luciferase takes part in its bioluminescence It has been reported to squirt lovely luminous clouds from its luminous organs (https://www.britannica.com/science/bioluminescence/The-range-and-variety-of-bioluminescent-organisms).Frank and Case, (1988) observed that it is bioluminescent due to its spew and photophores and the spectral sensitivity maxima of this species were at 400 and 500 nm and chromatic adaptation experiments indicated the presence of two visual pigments.Johnson *et al.* (https://doi.org/10.1515/9781400875689-032) reported that hot- and cold-water extracts of this shrimp displayed a luciferin-luciferase reaction. Herring (1981) however reported that the emission maxima of this species were at 455,457 and 462nm. It is also reported that the bioluminescent reaction takes place in this species when the oxidation of coelenterazine (substrate) with molecular oxygen is catalyzed by its luciferase, resulting in light of maximum intensity at 462 nm and the products CO2 and coelenteramide (https://patents.justia.com/patent/20150064731)

Oplophorus novaezeelandiae (= *Hoplophorus novaezealndiae*)

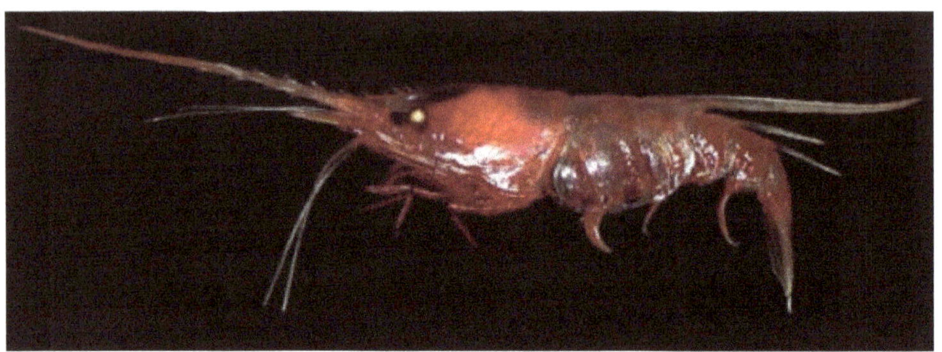

Image credit: Caroline A. Farrelly and Shane T. Ahyong (Applied for permission)

Common name (s): Not known

Global distribution: South Western Pacific Ocean; E. of New Zealand; Indian Ocean; South of Madagascar; South Atlantic;

Habitat: This pelagic; species has a depth range of 0 - 730 m.

Biology: In this species, rostrum is prominent. Spines are absent on the outer margin of antennal scale. Exopods of at least third maxillipeds and first legs are foliaceous and rigid. It grows to a maximum length of 10 cm.

Bioluminescence: In this species, its bioluminescence is associated with its luminous secretion discharged by its antennal glands or head glands (Nicol,1958).

Oplophorus spinosus (= *Hoplophorus grimaldi*)

Common name (s): Not known

Global distribution: Indian Ocean; Pacific Ocean: off Japan, Hawaii and Easter Island; subtropical Atlantic.

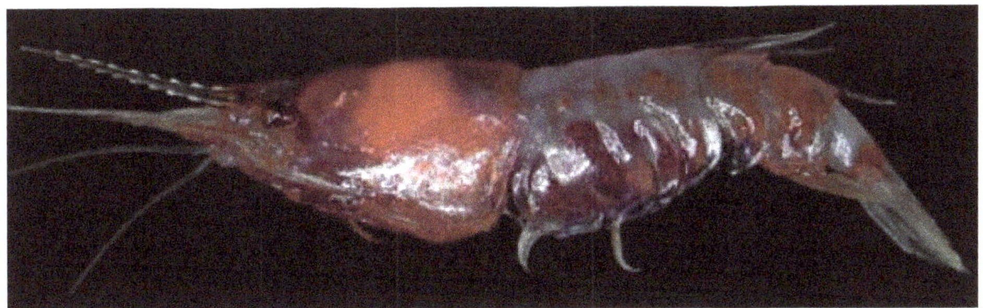

Image credit: Caroline A. Farrelly and Shane T. Ahyong, (Applied for permission)

Habitat: This mesopelagic species lives in deep midwater offshore. While it comes to the shallow waters (140 m) at night, it goes to deeper waters (750 m) during the day.

Biology: Body of this species is partly pinkish-red and partly transparent. Eyes are at least as wide as the eyestalk. Abdomen has no posterior mid-dorsal spine on its 2nd segment but the 5th abdominal segment which is longer than the 6th segment has a posterior mid-dorsal tooth, and the 5th segment is longer than the 6th segment. Telson is pointed posteriorly but it does not end with a spinose endpiece. It has no sharp tooth along the posterior end of the ventral margin of the carapace in adults.

Bioluminescence: Frank and Case, (1988) observed that its luminescence is due to its spew and photophores and the spectral sensitivity maxima of this species were at 400 and 500 nm and chromatic adaptation experiments indicated the presence of two visual pigments. Herring (1981) however reported that its emission maxima were 455-475nm.

Nowel *et al.* (1998) reported that two types of cuticular photophore have developed in this species in the third maxilliped and pleopods respectively. Photophores present in the third maxilliped consist of a unit comprising a single photocyte and associated pigment cells. The reflecting pigment cells contain white pigment and form an apical cap above the photocyte; and sheath cells contain red carotenoid pigment and form a light-absorbing layer around the photophore. The photophores located on the pleopods are compound structures comprising several photocytes. They also contain the same types of pigment cell as in the photophores of the maxilliped. These two mechanisms have been reported to optimize the use of the photophores in ventral camouflage; and they allow photophore rotation according to the shrimp's orientation so as to maintain the ventral direction of the luminescence.

Oplophorus typus

Common name (s): Not known

Global distribution: Indo-West-Pacific:

Image credit: Corbari L. - MNHN, Wikimedia Commons

Habitat: This bathypelagic species has a depth range of 200 - 1404 m. It is also an inhabitant of deep scattering layer from the west coast of India

Biology: Body of this species is robust and shell is hard. Abdominal somites III, IV and V are produced into overhanging spine posteriorly. Rostrum is short, with 7–9 dorsal and 4–7 ventral teeth.

Bioluminescence: This species is also known for its hepatic luminescence, photophores, and secretions (Poupin *et al.*, 1999). Herring (1981) reported that in this species, the emission of a luminous substance from its cephalothorax and emission maxima were at 456 and 457nm.

Systellaspis debilis

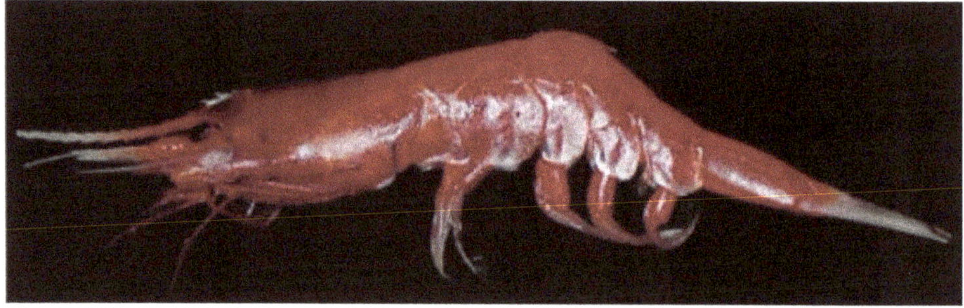

Image credit: Caroline A. Farrelly and Shane T. Ahyong, (Applied for permission)

Common name (s): Not known

Global distribution: Greenland, eastern and western Atlantic, South Africa and Indo-Pacific.

Habitat: This bathypelagic species has a depth range of 150–4594 m.

Biology: In this species, its carapace with rostrum is overreaching scaphocerite,.Ventral margin is with about ten teeth, and dorsal margin has 15 teeth. Antennal spine is absent. Abdomen is not dorsally carinate on all somites. Male pleopod 1 has endopod which is rounded and bilobed at apex. Male pleopod 2 is with appendix masculine which is little longer than appendix interna, and is rounded on distal portion. Body of the animal is in bright red coloration.

Bioluminescence: This species has been reported to discharge its luminous secretion from its antennal glands or head glands (Nicol, 1958). Frank and Case (1988) reported that it is a four photophore-bearing species and it has its sensitivity maxima at 400 and 500 nm. Chromatic adaptation experiments indicate the presence of two visual pigments in this species.

Systellaspis pellucida

Image credit: SEFSC Pascagoula Laboratory, Wikimedia Commons

Common name (s): Not known

Global distribution: Gulf of Mexico, Bahamas to Brazil, western Africa, and Indo-Pacific

Habitat: This bathypelagic species has a depth range of 101–3700 m.

Biology: In this carapace, its carapace with rostrum is overreaching scaphocerite. While the dorsal margin is with 9–11 teeth and 5–6 pre-orbital spines, ventral margin has 3–6 teeth. Strong antennal spine is present. Abdomen is not dorsally carinate on all somites. Somites 3–5 are with posteromesial tooth, the one of somite 3 is distinctly strong.

Bioluminescence: Poupin *et al.* (1999) reported that its luminescence is in the form of expulsion of a jet of bioluminescence and hepatic photophores. Herring (1981) observed that it has emission maxima at 455.456 and 458nm.

Family: Pandalidae

Chlorotocoides spinicauda

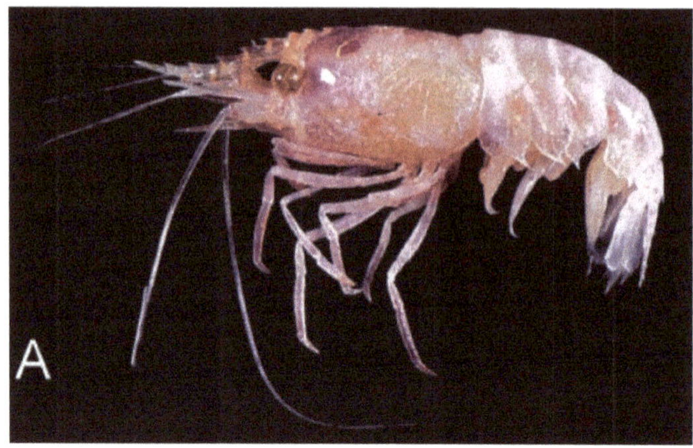

Image credit: Tin-Yam Chan, National Taiwan Ocean University/MNHN, Reproduced with permission

Common name (s): Not reported

Global distribution: Andaman and Maldive Islands; the South China Sea, Indonesia, and the Philippines

Habitat: It is presumably semi-pelagic or nekto-benthic with a depth range of 15–141 m.

Biology: Body coloration of this species is pinkish translucent and is covered with minute red dots. Rostrum has alternating, incomplete orange-pink bars. Eyes are dark brown. A distinctive yellowish ring is visible inside the anterodistal part of the carapace. Male grows up to a length of 9 mm.

Bioluminescence: The hepatic photophores of this species are believed to be the sources of the luminescence. These photophores do not contain paracrystalline material and they have a diffuse reflector which is separated from the tubules that make up the photocytes. Luminescence in the photophores of this species is probably intracellular (Herring,1981; Poupin *et al.*, 1999)

Heterocarpus dorsalis (= *Heterocarpus alphonsi*)

Common name (s): Madagascar nylon shrimp, caridean shrimp

Global distribution: Indo-Pacific: East Africa to the Pacific.

Habitat: This benthic species occurs in blue mud bottom at a depth range of 185 - 1400 m

Biology: The characteristic feature of this genus is the presence of a crest on third and fourth abdominal segments ending in a spine

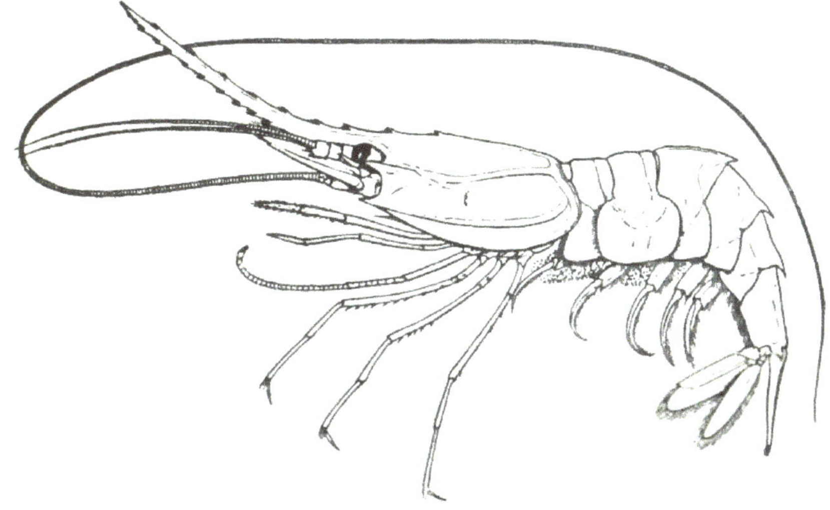

Image credit: Sir Ray Lankester, Wikimedia Commons

Bioluminescence: Its bioluminescence is largely due to its luminous secretion discharged by its antennal glands and head glands (Nicol,1958)

Heterocarpus ensifer

Image credit: NOAA, Wikimedia Commons

Common name (s): Caridean shrimp

Global distribution: Western and eastern Atlantic, western Indian Ocean, western and Central Pacific

Habitat: This bathypelagic species has a depth range of 140–950 m.

Biology: The characteristic feature of this genus is the presence of a crest on third and fourth abdominal segments ending in a spine

Bioluminescence: It emits blue light which is by vomiting" or "spitting" bioluminescent fluid into the water where they mix to generate light. The wavelength of emission maxima were at 455, 462, and 470 nm (Johnson *et al.*, 2012)

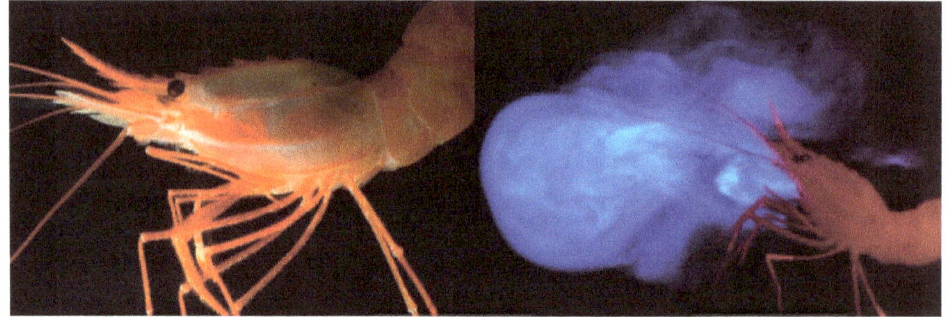

Heterocarpus ensifer **Close-up and after "Vomiting" or "Spitting" Bioluminescent Fluid**

Image credit: Wikiwand

Heterocarpus sibogae

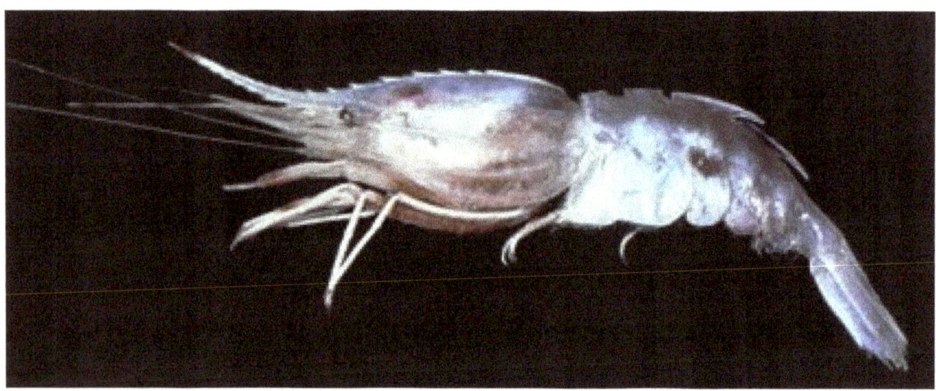

Image credit: MNHN and OFB [Ed]. 2003-2022. National inventory of natural heritage (INPN) (Applied for permission)

Common name (s): Mino nylon shrimp, little spewing shrimp

Global distribution: Indo-Pacific: Andaman Sea to Japan and Nazca Ridge.

Habitat: This benthic species is found on gray or green mud, fine sand or fine sand shells and pebbles at depths of 190 - 950 m

Biology: This bilaterally symmetrical species rely on drag powered swimming to move around. It grows to a maximum length of 11 cm.

Bioluminescence: This species has been reported to exhibit bioluminescence by emitting two streams of blue luminous secretions from the mouth when the animal is stimulated. The luminous cloud so discharged decayed rapidly and disappeared in few seconds. In laboratory conditions this species was found to release luminous secretions for a minimum period of 43 days under maintenance. In non-ovigerous females, the maximum emission time was found to be 634 sec. Males and non-ovigerous females displayed more or less a similar average emission time of about 30 sec. However, ovigerous females produced longer emission time which was about 103 sec. This is believed to be due to the possession of larger amounts of key chemical for bioluminescence *viz.* coelenterazine by the non-ovigerous females and males (Chan *et al.*, 2008)

Plesionika sp. (= *Parapandalus* sp.)

Common name (s): Caridean shrimp

Global distribution: Tropical and sub-tropical seas; Pacific Islands

Habitat: This deep-sea benthic species inhabits at depths of 500-1000m. It is commonly found at a depth of 690 m.

Biology: In this unidentified species exopods are present on first to fourth legs. Rostrum is at least twice the carapace length, and is closely and evenly toothed ventrally almost up to tip. Maximum length is 16 cm.

Bioluminescence: It is known to vomit blue light on the ocean floor. It is also a spew' variety in which the reagents are released into the water where they mix to generate light. The peak wavelengths of this light were found to be 470 and 455 nm. This animal spews light as a defense mechanism(Johnson *et al.*, 2012). Herring (1981) reported that the hepatic photophores of this species are believed to be the sources of the luminescence. These photophores do not contain paracrystalline material and they have a diffuse reflector which is separated from the tubules that make up the photocytes. Luminescence in the photophores of this species is probably intracellular (Herring,1981)

Thalassocaris crinita

Common name (s): Not reported

Global distribution: Indo-West pacific; Red Sea through the Indian Ocean to Indonesia; Philippines, Japan, and the Marshall Islands;

Habitat: It occurs commonly in shallow coastal waters near coral and other reef substrates at depths of 5-33 m. This mesopelagic species can be extremely seasonally abundant in upwelling areas such as the Arabian Sea where it is believed to be the most common micronekton species especially during the winter season

Biology: Body of this species is generally translucent and is covered with minute red dots. Eyes are dark brown. Rostrum is long. Females reach a maximum length of 3.7 mm.

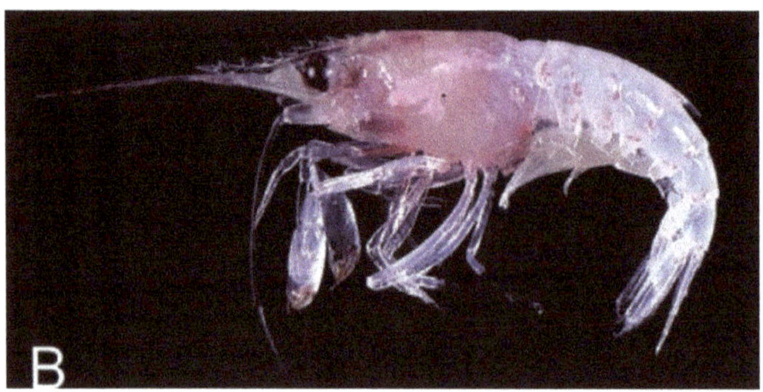

Image credit: Tin-Yam Chan, National Taiwan Ocean University/MNHN., Reproduced with permission

Thalassocaris lucida

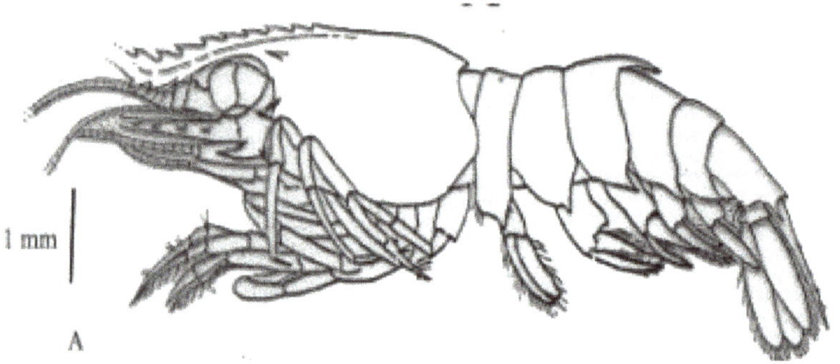

Common name (s): Not reported

Global distribution: Indo-West pacific; South Africa, eastern Indian Ocean, Japan, Marianas, and the Fiji Islands.

Habitat: It occurs commonly in shallow coastal waters and it may also be seen beyond continental shelf

Biology: Rostrum of this species is basally broadened into rather wide supraorbital eaves with subparallel margins. Telson is more than 4 times as long as wide. Second pair of pereoipods posseses no massive chelae. Individuals of this species have an average length of 2. 6 mm.

Bioluminescence in *Thalassocaris crinita* and *Thalassocaris lucida* : Herring (1981) and Poupin *et al.* (1999) reported that in these species, luminescence is in the form of expulsion of a jet of bioluminescence and hepatic photophores play an important role. These hepatic photophores do not contain paracrystalline material and they have a diffuse reflector which is separated from the tubules that make up the photocytes. Luminescence in the photophores of this species is probably intracellular.

Emmerson (2017) reported that these species can produce a blue bioluminescent secretion. Herring and Barnes (1976) reported on the presence of two pairs of light organs one close to the maxilla and one behind leg 5 in these species.Herring (1981) reported the emission maxima in these species as 450.456 and 460 nm.

Family: Pasiphaeidae

Glyphus marsupialis

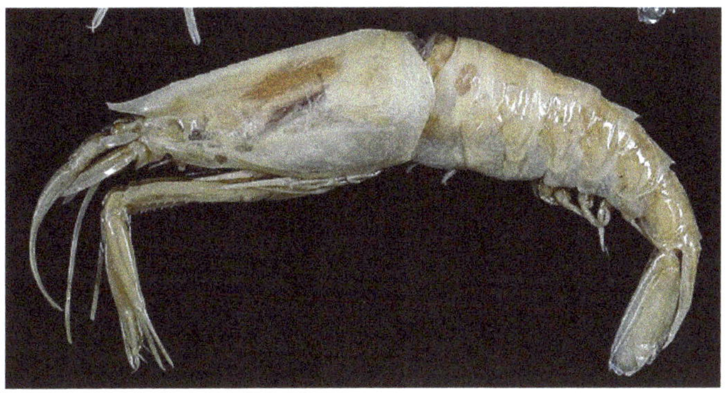

Image credit: Museum national d'Histoire naturelle, Wikimedia Commons

Common name (s): Kangaroo shrimp

Global distribution: Pacific, Indian and Atlantic Oceans

Habitat: This benthic or benthopelagic species inhabits waters over muddy sand bottoms of the continental slope at depths from 500 to 1230 m.

Biology: It is a large deep-sea shrimp and its body may exceed 20 cm. Cephalothoracic carapace of this species has a dorsal ridge and is strongly flexed. Abdomen of the females is swollen and therefore it is called as kangaroo shrimp.

Bioluminescence: The luminescence of this species is due to its liver extracts. The emission maxima of the displays of this species were in the range of 445-460 nm (Poupin *et al.*, 1999)

Family: Sergestidae

Allosergestes sargassi

Common name (s): Not known

Global distribution: Atlantic, the Mediterranean and Indo-Pacific.

Habitat: This pelagic species has a depth range of 0 - 2150 m. This species has been reported to depend on drag powered swimming to move around.

Biology: Carapace of this species has vertical frontal margin and minute "beak" on top.

Image credit: S.O.Guresen and O.Gonulal. (Applied for permission)

It has red chromatophores. Rostrum is short and triangular. Supraorbital spine, dermal photophore, and organ of Pesta are absent. Maxilliped 3 is longer than pereopod 3. Comb-like spinations are absent on the dactyl and distal half of propodus of maxilliped 3. Telson is 4.0 times as long as wide. Eyestalk is 2.1 times as long as wide; and cornea is 0.9 times as long as wide, 0.6 times as long and 1.3 times as wide as eyestalk.

Bioluminescence: Sergestes is the only genus characterised by the presence of organs of Pesta. In this species, both anterolateral and posterolateral organs of Pesta are spheroid. It has distinct posterolateral light organ arrays: which are trilobed. Across considered emission intensities, maximum possible sighting distance has been reported to be 5.45 m for this species. (Schweikert *et al.*, 2020)

Deosergestes corniculum

Common name (s): Not known

Global distribution: Temperate-subtropical North and South Atlantic

Habitat: This species has been reported to live at a depth of 35 m during night and 170 m during day time

Biology: Carapace of this species is bluntly rounded and is rarely with minute tooth on anterior margin. Further its carapace is 2.7 times as long as high and 0.44 times as long as abdomen. While supraorbital spine is absent, hepatic spine is moderately developed. Telson is 4.2 times as long as wide. Eyestalk is 2.1 times as long as wide. Cornea is as long as wide; and it is 0.7 times as long and 1.4 times as wide as eyestalk.

Bioluminescence: In this species, its anterolateral organs of Pesta are lobed and posterolateral organs are fringed. The peak wavelength of light emission for this species was at 469 nm (Schweikert *et al.*, 2020).

Deosergestes henseni

Common name (s): Not known

Global distribution: NW Africa, Straits of Florida and Northern Gulf of Mexico

Habitat: The individuals of this species occur at depths of 150–1,500 m.

Biology: Carapace is 2.4 times as long as high and 0.48 times as long as abdomen. Supraorbital and hepatic spines are well developed. Abdomen with somite VI is 1.6 times as long as high and 1.2 times as long as telson. Telson is 4.8 times as long as wide. Eyestalk is 1.8 times as long as wide. Cornea is as long as wide, 0.6 times as long and 1.2 times as wide as eyestalk.

Bioluminescence: In this species, both anterolateral and posterolateral organs of pesta are spheroid. It has distinct posterolateral light organ arrays which are fringed. Across considered emission intensities, maximum possible sighting distance has been reported to be 7.57 m for this species. Across all model conditions, this species had the greatest sighting distance estimates of intraspecific bioluminescence, followed in order by *Parasergestes armatus* and *Allosergestes sargassi* (Schweikert *et al.*, 2020)

Eusergestes arcticus

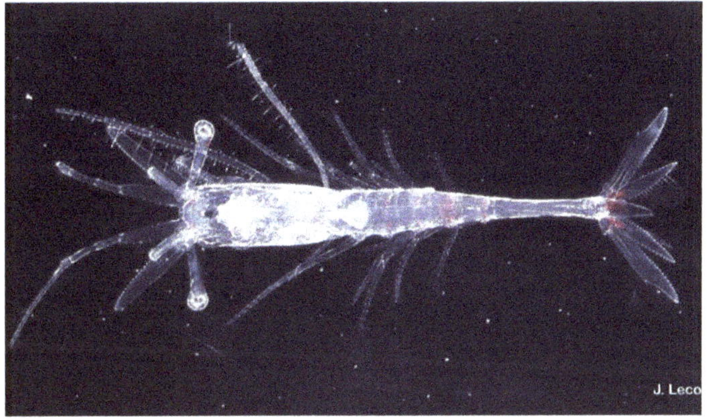

Image credit: Tamara Frank, Reproduced with permission

Common name (s): Not known

Global distribution: Indo-Pacific, Atlantic, Mediterranean, Arctic and Antarctic:

Habitat: This free-swimming, mesopelagic species occurs at a depth range of 180 - 4500 m. Its daytime depth is 600 - 960 m.

Biology: In this species, its carapace is laterally compressed and is more than half the length of the abdomen, excluding the telson. Rostrum is very short reaching only as far as the basal articulation of the eyestalks. Eyes are long and slender and cornea is round and wider than the stalk. First pair of pereiopods are not chelate and is reaching slightly beyond the middle of the antennular peduncle. Second and third pairs possess a minute but perfectly formed chela. Telson is shorter than both uropods. Walls of the carapace are transparent, with a few small scarlet red chromatophores. Its black stomach and scarlet hepatic and cardiac regions show through very distinctly. Length of the individuals is up to 65 mm.

Bioluminescence: In this species, the anterolateral organs lobed and posterolateral organs of pesta are fringed. All the species of *Sergestes* have been reported to possess internal photophores (light organs) which are functioning in countershading. Its downward directed bioluminescence has a spectral emission, angular distribution and irradiance which are all consistent with a camouflage function. The ability of this shrimp to alter body colouration is to adapt to its prevailing light regime (Vestheim and Kaartvedt,2009)

Eusergestes similis (= *Sergestes similis*)

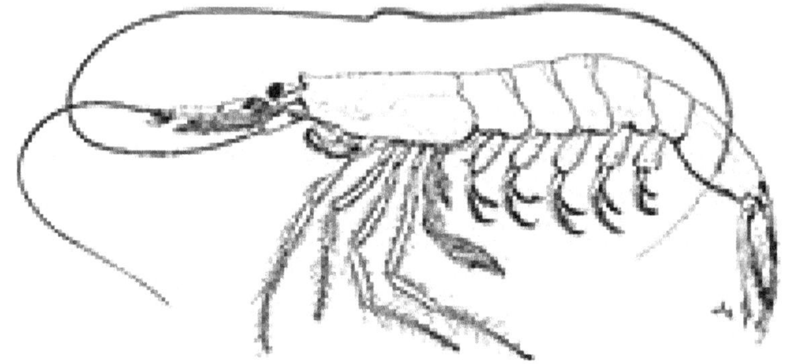

Image credit: FAO

Common name (s): Common midwater shrimp.

Global distribution: Arctic, Indo-Pacific and Southeast Atlantic.

Habitat: Adults of this species are mesopelagic, and are mostly oceanic. It may also tolerate brackish conditions and is also found in mangrove vegetate area

Biology: Body of this species is slender. Carapace and overall cuticle surface is smooth. Rostrum is short, and is not extending beyond eyes. Tip of rostrum is formed into an acute spine and a small spine is seen on either side of rostrum. Eye is small and cornea is well developed. Eye stalk is long. Atennule peduncle is very long, and is exceeding atennule scale. Supraorbital and hepatic spines are seen. Adults have a total length of 40-54 mm.

Bioluminescence: The light organs of this species are large, unlensed modifications of the hepatopancreas known as "organs of Pesta" (luminescent modified areas of the gastrohepatic glands) found within the cephalothorax.

Further, both anterolateral and posterolateral organs of pesta are spheroid. In this species, the peak wavelength of light emission was at 469 nm (Schweikert *et al.*, 2020). Herring (1981) reported that the hepatic photophores of this species possess fluorescent paracrystalline platelets which are believed to be the sources of the luminescence. The cells containing these crystals are contiguous with cells filled with lipid droplets which act as a diffuse reflector. Luminescence in the photophores of this species is probably intracellular((Herring,1981)

Parasergestes armatus

Image credit: WoRMS (CC)

Common name (s): Not reported

Global distribution: Straits of Florida, Northern Gulf of Mexico

Habitat: This species has been reported to occur at depths of 150–1,500 m.

Biology: Carapace of this species is 2.6 times as long as high and 0.50 times as long as abdomen. Supraorbital tooth is present. Abdomen with somite VI is 1.7 times as long as high and 1.4 times as long as telson. Telson is 3.3 times as long as wide. Eyestalk is 2.3 times as long as wide. Cornea is 1.1 times as long as wide, 0.7 times as long and 1.4 times as wide as eyestalk

Bioluminescence: In this species, both anterolateral and posterolateral organs of pesta are spheroid. It has distinct posterolateral light organ arrays which are bilobed. Across considered emission intensities, maximum possible sighting distance-6.18m (Schweikert *et al.*, 2020)

3.7. Mollusks

Among the luminous marine mollusks, the class Bivalvia has only the

Pholas. The luminous gastropods include the genera *Hinea, Kaloplocamus, Phyllirrhoe Planaxis* and *Plocamopherus.* Of these, the species of *Phyllirrhoe* are planktonic and are dealt with here. Unlike the Bivalvia and Gastropoda, the class Cephaopoda has a number of luminous genera which are characterized by the presence of light-producing organs, also known as photophores.The cephalopod photophores may range from simple groups of photogenic cells, to organs with photogenic cells which are surrounded by reflectors, lenses, light guides, color filters, and muscles. Further, these luminescent organs vary from very small, simple photophores of less than 0.2 mm in dia. (*e.g.* mesopelagic squid *Abralia trigonura*) to larger, more complex organs from squid of the family Sepiolidae. Complex photophores, also referred to as light organs, are found in the sepiolid squid *Euprymna scolopes* and in a number of other species in the families Sepiolidae and Loliginidae. Light organs of these animals are known to adjust color, intensity, and angular distribution of light produced from within. On the other hand, in oceanic cephalopods, the photophores emit intrinsic luminescence (autogenic) with light coming from their own photocytes. Further, these cells are also able to display different spectra of light. In contrast, photophores of several neritic cephalopods possess extrinsic luminescence (bacteriogenic), with light produced by the photogenic bacteria housed in a specialized light organ complex within the mantle cavity of their host. Light organs of similar morphology are also seen in oceanic teuthid squids (epipelagic and mesopelagic), such as *Chiroteuthis* spp., *Chiropsis* spp. and *Taningia* spp. However, these anal light organs do not house luminous bacteria. It is also reported that in shallow-water cephalopods, visual behaviors are the major components of intra-interspecific communication. Further, the meso- and bathypelagic cephalopods possess limited visual communication other than bioluminescence and this is due to the reduced illumination at deep-sea (Haddock, http://photobiology.info/Haddock.html; https://biolum.eemb.ucsb.edu/organism/)

3.7.1. Bivalve

Class: Bivalvia or Pelecypoda

Pholas dactylus

Image credit: Hectonichus, Wikimedia Commons

Common name (s): Common piddock

Global distribution: North Atlantic and the Mediterranean Sea

Habitat: This coastal species is found from the lower shore to the shallow sublittoral zone up to depth of 35 m.

Biology: It has been reported to bore into a wide range of substrata including various soft rocks such as chalk and sandstone and very occasionally in waterlogged wood. It is approximately elliptical in outline with a beaked anterior end which is up to 12 cm long. Shell is thin and brittle with a sculpture of concentric ridges and radiating lines. Shell is dull white or grey in color and the periostracum is yellowish. Siphons are joined and are at least one to two times the length of the shell.

Bioluminescence: This species is one of the longest-known and best studied luminous mollusc. The pioneering experiments made in this species paved the way for the existence of a luciferin+luciferase reaction (Haddock *et al.*, 2020). The presence of the oxidised form of coelenterazine *viz.* dehydrocoelenterazine in this species was identified as the luciferin required in the luminescence of this clam (Mirza and Oba,2021)

3.7.2. Gastropods

Class: Gastropoda

Hinea brasiliana

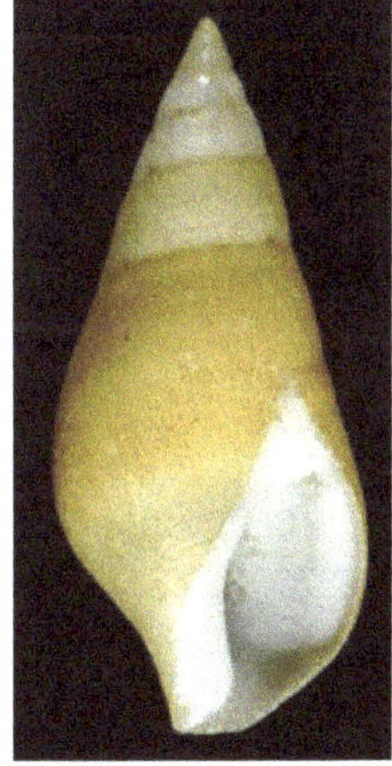

Image credit: WoRMS. (CC/ NC) (Applied for permission)

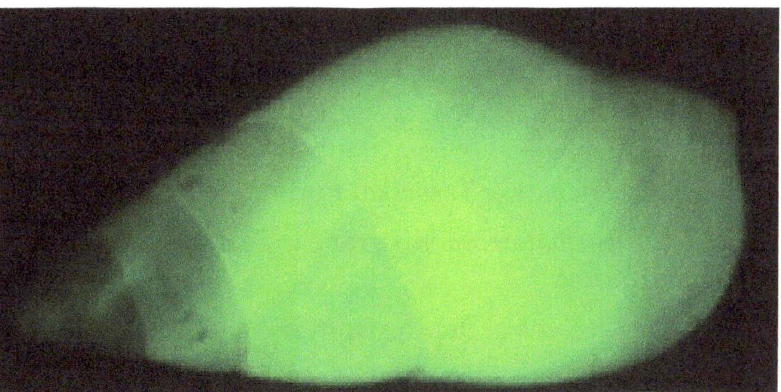

Image credit: N.Wilson, Australian Geographic (Applied for permission)

Common name (s): Yellow-coated clusterwink,

Global distribution: It is native to New Zealand and southeastern Australia.

Habitat: It is found in the littoral zone of rocky shores among boulders and rubble.

Biology: Shell of is species is thick and heavy and grows to a length of 21 mm. It is narrowly conical, either smooth or with shallow grooves between its six spiral whorls. Aperture is small and constricted by a callus and the columella, There is a horny operculum. Colour of the shell is white.

Bioluminescence: This species has been reported to emit bright flashes of green light (https://www.australiangeographic.com.au/topics/science-environment/2018/08/bioluminescent-beauties-meet-australias-sparkling-species/). It is also reported that reported that when disturbed, this species emits a series of short flashes of bluish-green light which is produced from its mantle tissue and it shines through the pale translucent shell (https://en.wikipedia.org/wiki/Hinea_brasiliana)

Kaloplocamus ramosus

Image credit: Seascapeza, Wikimedia Commons

Common name (s): Tasselled nudibranch

Global distribution: Mediterranean Sea, south-eastern Australia,Japan, Hong Kong and Korea

Habitat: It lives at depths from 25–400 m

Biology: It is a very well-camouflaged nudibranch and it cannot be noticed underwater without using a torch. Body of this species is pale with variable amounts of reddish-pink pigmentation and is covered with raised white spots. It has large perfoliate rhinophores, which are usually pinkish in color

Bioluminescence: In this species, the luminescence is intrinsic which means that the production of light is dependent on the animal's own biochemical processes and there is no symbiotic association with bacteria. Further, in this species, the luminescence is intracellular and its chemical reaction takes place within luminescent cells or photocytes (Valle's and Gosliner,2006).

Planaxis sp.

Image credit: Wikipedia

Common name (s): Sea snail, periwinkle

Global distribution: Indian Ocean

Habitat: The species of Planaxis occur in abundances along the shore, in the upper and middle zones of intertidal environments

Biology: Shell of the species of *Planaxis* is solid, thick, spotted or tessellated with chocolate or black and white. It is spirally uniformly grooved. There are five whorls. Body whorl is the largest and is faintly angular. Aperture is subtrigonal, Columella is straight, and white. Outer lip is thin and wavy corresponding to groove on body-whorl.

Bioluminescence: It is a bioluminescent species (Haddock *et al.,* 2010). No other information is available.

Plocamopherus ceylonicus

Common name (s): Not known

Global distribution: Western Pacific: Hong Kong, New Caledonia and Vietnam; Sri Lanka

Habitat: It is a benthic species commonly occurring in areas of rocky reef and seagrass beds

Biology: It is a is a moderately sized, translucent-gray animal attaining a maximum length of 55 mm. Brown patches and speckling with orange-yellow spots are seen around the edge of the foot and on the mantle. Posterior tip of the foot extends into a long tapering 'tail'.

Image credit: Rudman, W.B., Sea Slug Forum. Australian Museum, (Applied for permission)

Bioluminescence: In this species, the four papillae surrounding the gills are tipped by a large rounded pinkish knob which flashes light when disturbed. Further, there are often flashes of light from the sides of the body (http://www.seaslugforum.net/showall/plocceyl). Valle's and Gosliner (2006) reported that in the species of *Plocamopherus* the luminescence is intrinsic which means that the production of light is dependent on the animal's own biochemical processes and there is no symbiotic association with bacteria. Further, in this species, the luminescence is extracellular and its chemical reaction results from the discharge of the luminous chemicals outside the cell where the chemical reaction occurs.

Plocamopherus imperialis

Common name (s): Imperial plocamopherus

Global distribution: Tropical to subtropical; Indo-West Pacific: Indonesia to New Zealand.

Habitat: It is a benthic species and occurs from intertidal and sub-tidal waters to a depth of at least 10 m.

Biology: This species is recognised by its orange or reddish body with brown or black speckles. Tail of the animal is pointed and flattens into a paddle shape when the animal "swims". A series of large compound papillae arise from the oral veil and the sides of the mantle. It grows to about 100 mm in length.

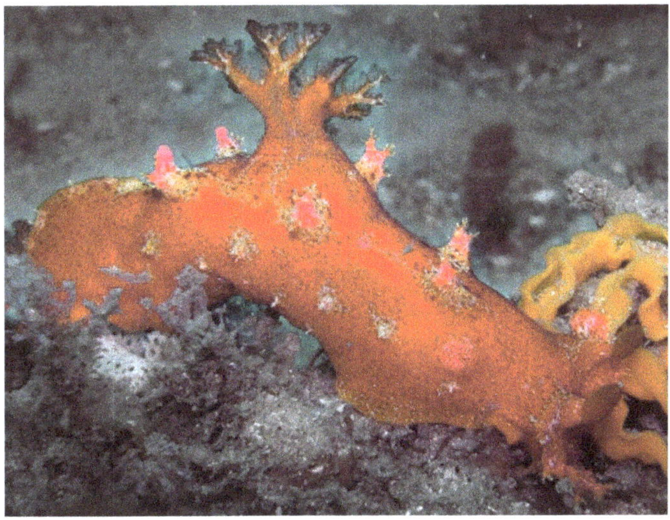

Image credit: Sylke Rohrlach, Wikimedia Commons

Bioluminescence: Valle's and Gosliner (2006) reported that in the species of *Plocamopherus* the luminescence is intrinsic which means that the production of light is dependent on the animal's own biochemical processes and there is no symbiotic association with bacteria. Further, in this species, the luminescence is extracellular and its chemical reaction results from the discharge of the luminous chemicals outside the cell where the chemical reaction occurs.Anon. (https://www.surg.org.au/species/nudibranchs-and-sea-slugs/polyceridae/plocamopherus/imperialis) reported that some of the papillae arising from the oral veil and the sides of the mantle of this species often have a large bulb at the tip, which, when the animal is disturbed, luminesce, producing flashes of light

Plocamopherus maderae

Image credit: Bernard Picton, Wikimedia Commons

Common name (s): Not reported

Global distribution: This species was originally described from Madeira. (an autonomous archipelago of Portugal). It has subsequently been reported from the Canary Islands and the Cape Verde Islands

Habitat: This nudibranch is normally nocturnal, hiding beneath boulders in shallow water during daytime.

Biology: This nudibranch is orange or yellow-orange in color, with numerous small dark brown spots and few, larger, orange spots and patches. Its mantle edge has three pairs of branched papillae. Tail is long, flattened and it has a dorsal crest. Gills are large, sparsely branched and are held erect. It attains a maximum size of about 50 mm.

Bioluminescence: In this species, the papillae located behind the branchial plume bear rounded bioluminescent organs (https://en.wikipedia.org/wiki/Plocamopherus_maderae).Valle's and Gosliner (2006) reported that in the species of *Plocamopherus* the luminescence is intrinsic which means that the production of light is dependent on the animal's own biochemical processes and there is no symbiotic association with bacteria. Further, in this species, the luminescence is extracellular and its chemical reaction results from the discharge of the luminous chemicals outside the cell where the chemical reaction occurs.

Plocamopherus tilesii

Image credit: Sylke Rohrlach, Wikimedia commons

Common name (s): Tile's sea slug

Global distribution: Indo-West Pacific; Japan, China and New South Wales

Habitat: This shell-less marine gastropod is a benthic species and is found in reef flat shallows with soft sediment at a depth of 13 m

Biology: It has a large flat tail and broad flat head. It attains a maximum length of 6 cm.

Bioluminescence: Valle's and Gosliner (2006) reported that in the species of *Plocamopherus* the luminescence is intrinsic which means that the production of light is dependent on the animal's own biochemical processes and there is no symbiotic association with bacteria. Further, in this species, the luminescence is extracellular and its chemical reaction results from the discharge of the luminous chemicals outside the cell where the chemical reaction occurs.

3.7.3. Cephalopods

Class: Cephalopoda

Abralia veranyi

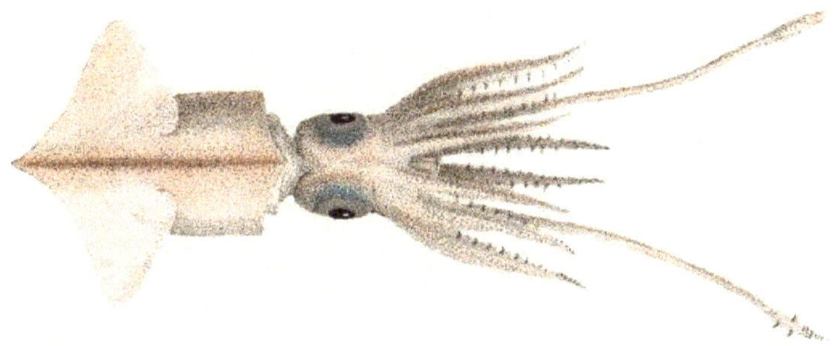

Image credit: Vérany, Jean Baptiste, Wikimedia Commons

Common name (s): Eye-flash squid, Verany's enope squid or midwater squid

Global distribution: Atlantic Ocean and the Mediterranean Sea.

Habitat: It undergoes a daily vertical migration from deep waters to near the surface.

Biology: It is a small species with a mantle length of about 4 cm. It has a fin, four pairs of arms and two long tentacles. Tips of the arms bear four rows of suckers and there are three hooks on the club of each tentacle. Male has the fourth arm on the left modified into a hectocotylus.

Bioluminescence: The underside of this squid bears about 550 light-producing organs called photophores. which are arranged in transverse rows each consisting of 4 to 6 large ones with many small ones in between. There are two large and three medium-sized photophores below each of the large eyes (https://en.wikipedia.org/wiki/Midwater_squid).In this species, its bioluminescence is largely due to its ventral photophores. It was found to be spontaneously luminescent when brought into the dark room. Luminescence was mostly of relatively constant intensity and any changes that did take place were gradual. However, any manipulation of the squid was liable to induce a very rapid and bright flash from one or more of the subocular photophores (usually the large posterior organ). Further the temperature variations showed difference in the pattern of emission of light in this species. For example at 11°C the spectrum was unimodal, peaking at 490 nm with a narrow bandwidth. At the higher temperature a shoulder at about 440 nm appeared in the emission spectrum. The short, bright flashes from the subocular photophores had a broader bandwidth and a shorter wavelength emission maximum (475 nm) than that of the ventral photophores (Herring *et al.*, 1992)

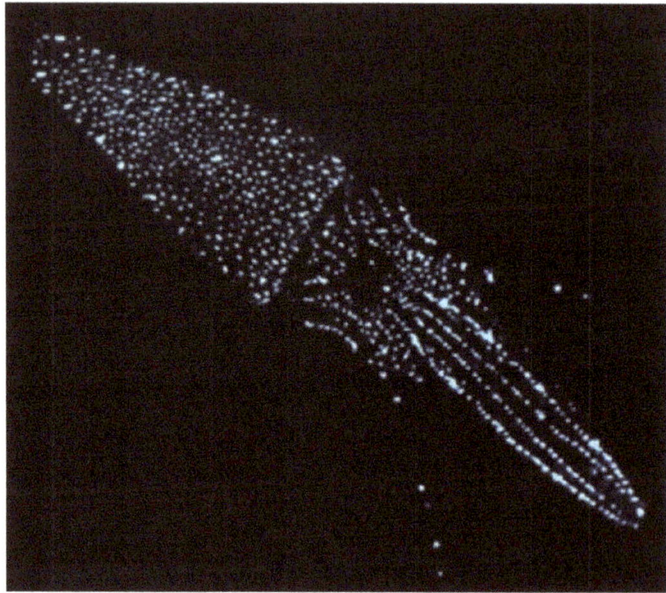

Glowing Photophores

Image credit: Smithsonian, Ocean Find Your Blue. (Applied for permission)

Abraliopsis (Abraliopsis) morisii

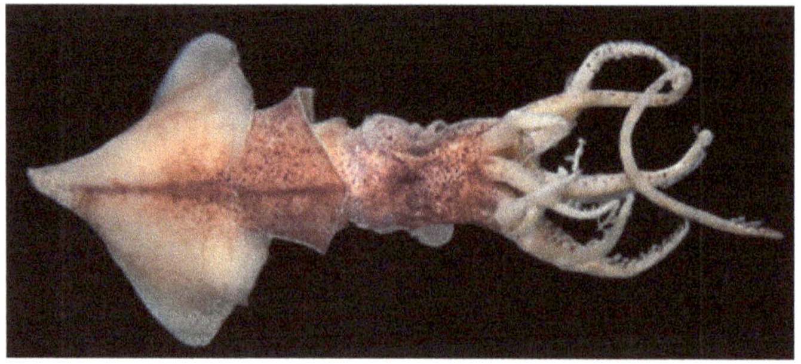

Image credit: Daniel J. Drew - Gall L, Wikimedia Commons

Common name (s): Pfeffer's firefly squid

Global distribution: Tropical to warm temperate waters in the Atlantic Ocean, including the Gulf of Mexico and the Mediterranean Sea

Habitat: It can be found in the epipelagic and mesopelagic zones at depths of between 0 and 3660 m.

Biology: It attains a maximum length of 7 cm.

Bioluminescence: In this species, there are integumental photophores which are scattered randomly and five ocular photophores. It is reported that its arms IV are with three terminal light organs. There are two kinds of photophores (large, spherical, dark; and small, spherical, translucent) in about nine indistinct rows on ventral surface of head. Further ventral periphery of eyes is with five photophores in single row; and there are 13 small integumentary photophores around eyelids (https://www.rawpixel.com/image/547488/bioluminescent-squid-vintage-poster).

Bathypolypus arcticus

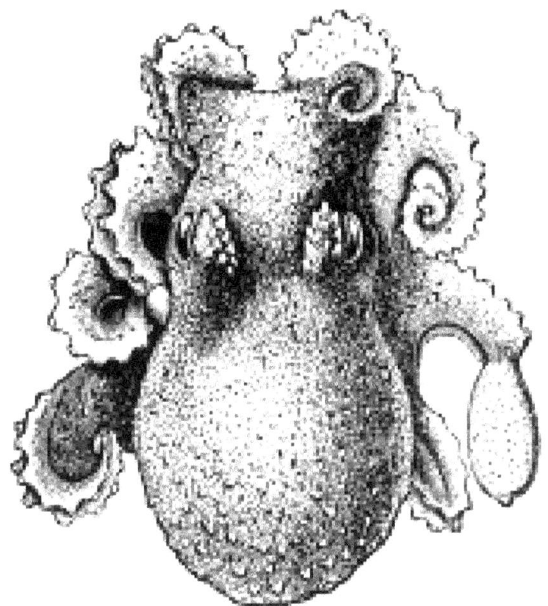

Image credit: Wikipedia

Common name (s): North Atlantic octopus, spoonarm octopus

Global distribution: Atlantic Ocean and the Arctic.

Habitat: This deep sea octopus that is typically found at depths from 200-600 m though its usual depth range is 14 - 2700 m.

Biology: It is a small, short-armed octopus reaching a maximum length of 10 cm and weight of 400g.Mantle is globular and is nearly as wide as it is long. Surface of mantle, head, arms and web is covered in warts, especially around the eyes. Arms are short and in an irregular order with two rows of small suckers.

Bioluminescence: It is a luminous species (https://www.science.org/content/article/vomiting-shrimp-and-other-deep-sea-creatures-light-ocean-floor)

Dosidicus gigas

Image credit: NOAA/MBARI, Wikimedia Commons

Common name (s): Humboldt squid, jumbo squid or jumbo flying squid

Global distribution: Eastern Pacific Ocean.

Habitat: It is most commonly found at depths of 200 to 700 m

Biology: This squid typically reaches a mantle length of 1.5 m. and weight up to 50 kg. Its two tentacles bear 100–200 suckers. Its nickname "red devil" comes from the fact that when it is caught by fisherman and brought to the surface, this squid turn a bright red color.

Bioluminescence: In this species, a blue bioluminescent light is emitted from an array of photophores on its body. These photophores are small, ovoid rice-like granules that are embedded in the muscle all over the squid on the mantle, fins, head, arms and tentacles. Like the bioluminescent clam, Pholas dactylus and flying squid, *Sthenoteuthis oualaniensis*, the oxidised form of coelenterazine vi. dehydrocoelenterazine is utilised in the light emission of Humboldt squid (Mirza and Oba,2021)

Eledonella pygmaea

Image credit: Ewald Rübsamen - Die Cephalopoden Carl Chun, Wikimedia Commons

*Common name (s):*Not known

Global distribution: Tropical to warm temperate waters in the Atlantic Ocean, including the Gulf of Mexico and the Mediterranean Sea.

Habitat: It can be found in the epipelagic and mesopelagic zones

Biology: In this species, eyes are small, elliptical, and are widely separated, There are three suckers in middle part of 3rd right arm which is enlarged in adult males. Color of the animal is mostly red or reddish; and it attains a maximum length of 25 cm.

Bioluminescence: The species of this genus have been reported to possess a greenish-yellow ring around their mouth which is only periodically luminous (Haddock *et al.*, 2010). It is also reported that the females of this species possess a peculiar circumoral light organ which luminesced brilliantly when treated with H2O2 (https://scholarspace.manoa.hawaii.edu/items/ab72894e-f208-4bc6-93c2-43c72d0791fc).

Euprymna scolopes

Common name (s): Hawaiian bobtail squid

Global distribution: Central Pacific Ocean; off the Hawaiian Islands and Midway Island

Habitat: It occurs in shallow coastal waters

Biology: It is one of the smallest and slimmest sepiolid squids. Males have slightly larger suckers than females, with thinner posterior mantles. Both sexes have a pair of unique paddle shaped fins It grows to a maximum length of 30 mm in mantle length and adults weigh up to 2.8 g only

Image credit: Margaret McFall-Ngai, Wikimedia Commons

Bioluminescence: A feature unique to this species is its large glandular, bilobed and bioluminescent light organ which is present inside the squid's mantle cavity. This organ, which functions through its interaction with its symbiotic partner Vibrio fischeri, provides light (https://animaldiversity.org/accounts/Euprymna_scolopes/). As the down-welling light increases or decreases, the squid is able to adjust luminescence accordingly, even over multiple cycles of light intensity (https://en.wikipedia.org/wiki/Euprymna_scolopes). Wei and Young (1989) reported that most individulas of this species that hatched into seawater containing, or previously exposed to adults produced light within 24 h. Access to an initial inoculum of free-living, luminous bacteria seems to be critical for establishing a successful symbiosis.

Euprymna tasmanica

Image credit: Mark Norman/Museum Victoria, Wikimedia Commons

Common name (s): Southern bobtail squid or southern dumpling squid

Global distribution: Southern Indo-Pacific

Habitat: It lives on sand and mud areas, in association with seagrass beds up to depths of about 80m. It can glue sand grains to its upper body to aid camouflage with its environment.

Biology: In this species, each arm possesses four rows of suckers. Mature males possess enlarged suckers on the inner and outer rows or arms 2-4.

Bioluminescence: In this species, In this species, a saddle-shaped bacterial light organ is present inside the mantle cavity (https://australian.museum/learn/animals/molluscs/southern-bobtail-squid-euprymna-tasmanica-pfeffer-1884/)

Heteroteuthis dispar

Image credit: Consejo Superior de Investigaciones Cientificas (Applied for permission)

Common name (s): Odd bobtail or bobtail squid

Global distribution: North Atlantic Ocean and the Mediterranean Sea.

Habitat: This species is found in the mesopelagic zone of the ocean at depths down to 1600 metres.

Biology: In this species, there are five pairs of appendages. Arms 1 and 2 are fused at the base and each is a hectocotylus in the male. Arms 3 and 4 are longer than the others. In the female, arms 1 and 2 have no suckers on the distal ends. Tentacles are retractable, slender and whip-like, with a short club with eight rows of microscopically small suckers

Bioluminescence: This bobtail squid has a symbiotic relationship with bioluminescent bacteria that grow in a large light organ on its belly. It has been reported to squirt the bioluminescent bacteria (bioluminescent clouds) into the face of an attacker instead of its ink (https://news.globallandscapesforum.org/57739/bioluminescence-it-requires-a-whole-new-way-of-seeing/

Inioteuthis japonica (= *Inioteuthis inioteuthis*)

Common name (s): Not known

Global distribution: Northwest Pacific: from Japan to China.

Habitat: It is a demersal species.

Biology: It grows to a maximum length of only 2 cm

Bioluminescence: This species has been reported to emit a cobaltish light from a great luminous organ which is situated in the mantle cavity near the ink-bag. Anatomical observations of this luminous organ have shown similarities with that of Sepiola (Berry,1920)

Iridoteuthis lophia

Common name (s): Not known; new species

Global distribution: New Zealand

Habitat: This species lives at a depth range of 143– 732 m

Biology: Head of this species is deep, and broad, Eyes are large, and bulbous and occupy most of the sides of the head. Arms 1, 2 and 4 are similar in length in both sexes. Arms are relatively longer in males than females. Tentacles ae slender; and stalks are naked and semicircular in cross section.

Bioluminescence: Bioluminescence of this species is largely similar to *I.merlini* (https://australian.museum/blog/amri-news/fire-shooting-butterfly-bobtail-named-in-honour-of-professor-merlin-crossley/)

Reid (2021) reported that its visceral photophore which is ventral in position is associated with its ink sac. Further this photophore is distinct, round, bulbous and iiridescent purple in preserved specimens.

Iridoteuthis merlini

Common name (s): Butterfly bobtail squid

Global distribution: endemic to the open ocean off New Zealand as well as eastern and south-eastern Australia.

Habitat: Adults mainly occur in soft-sediments, reef and fringe habitats. It is referred to as 'butterfly bobtail squid' because of the way it swims by flapping its round fins,

Biology: Individuals of this species have silvery sides, colourful shields on their underside and have huge round eyes that take up a large proportion of

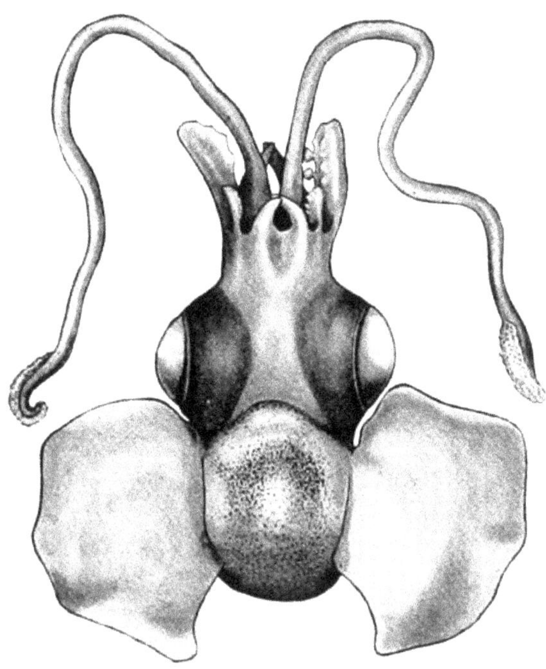

***Iridoteuthis* sp.**

Image credit: H. V. Poor, Wikimedia Commons

the sides of their heads. Males are with well developed keel on arms 3; and arms 4 with narrow keel. Females are with keels on all arms. It is also worthy of mention here that *Iridoteuthis* and *Stoloteuthis* species are superficially quite similar. Careful observation of the *Stoloteuthis maoria* type specimens revealed that the type specimens comprised not only two species, but represented two genera. The males are the 'real' *Stoloteuthis maoria*, and the females belonged to an undescribed species, now formally described and named *Iridoteuthis merlini*. (https://australian.museum/blog/amri-news/fire-shooting-butterfly-bobtail-named-in-honour-of-professor-merlin-crossley/)

Bioluminescence: It has been reported to emit sparkling light from a large light organ, or photophore which is filled with luminescent bacteria. It can also squirt out bioluminescent bacteria to form a light cloud, otherwise called 'fire shooting' (https://australian.museum/blog/amri-news/fire-shooting-butterfly-bobtail-named-in-honour-of-professor-merlin-crossley/)

Japetella diaphana

Common name (s): Diaphanous pelagic octopod

Global distribution: It has worldwide distribution in tropical and subtropical waters. It is common in Eastern tropical Pacific Ocean

Image credit: Ewald Rübsamen, Wikimedia Commons

Habitat: This pelagic species inhabits the oxygen minimum zone at a depth range of 200 - 4000 m.

Biology: The arms are also very short, no longer than the mantle length. Total length may be around 12 cm. Body of this species is small, gelatinous, and semi-transparent to orange in color with large, prominent eyes and eight short arms. Arm length is less than mantle length. Mature males lack hectocotylized arm, but do possess enlarged distal suckers on the third right arm. It's mantle length and total length are 8.5 cm and 16 cm respectively

Bioluminescence: Mature females of this species have a greenish-yellow ring-shaped light organ around the mouth which is only periodically luminous (Haddock *et al.*, 2010; http://dsg.mbari.org/dsg/view/concept/Japetella%20 diaphana).It is also reported that it has a bioluminescent, yellow ring which starts at the base of its tentacles, of which all are of the same size. The female's bioluminescent ring may be of a specific wavelength (https://en.wikipedia.org/wiki/Japetella)

Leachia sp.

Common name (s): Cockatoo squid, glass squid or cranchiid squid

Global distribution: It is found distributed in all open oceans around the world.

Habitat: It occurs in surface and midwater depths of oceans

Biology: It is characterized by a swollen body and short arms, possess two rows of suckers or hooks. Third arm pair is normally enlarged. Eye morphology is ranging from large and circular to telescopic and stalked. A large, fluid-filled chamber containing ammonia solution is used to aid buoyancy.

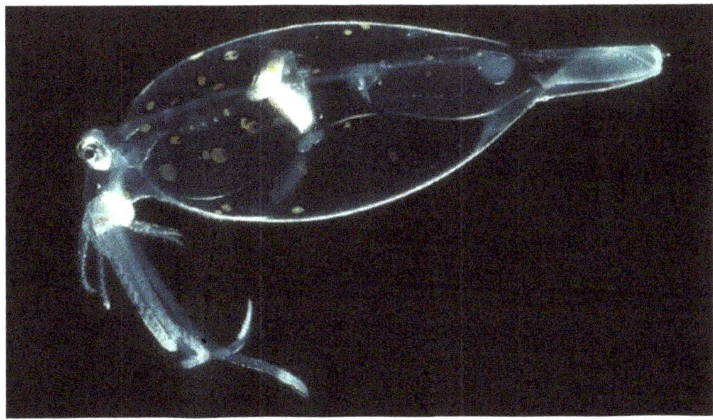

Image credit: Smithsonian, Ocean Find Your Blue. (Applied for permission)

Bioluminescence: It possess light organs on the undersides of its eyes, Its ocular photophores contain lenses and a thickened corneal layer, besides the gland cells (http://ecoursesonline.iasri.res.in/mod/page/view.php?id=42450)

Loliolus affinis

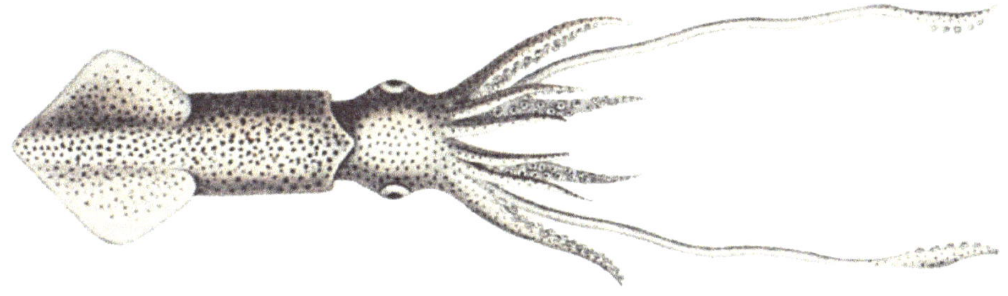

***Loliolus* sp.**

Image credit: Wikiwand

Common name (s): Steenstrup's bay squid or little Indian squid

Global distribution: Indo-West Pacific.

Habitat: It is a demersal species

Biology: It is a small animal growing to a length of only 4 cm. Mantle is slightly dorsoventrally flattened and it tapers very gradually posteriorly to a blunt rounded tip. Fins are large, heart shaped, and their posterior margins are straight. Arms are short. Left ventral arm of males is hectocotylised. Tentacles are relatively short, and their stalks are naked and with small, slightly expanded clubs.

Bioluminescence: In this species, its light emission is due to the presence of luminescent bacteria from its photophores (Anderson *et al.*, 2013)

Octopoteuthis deletron

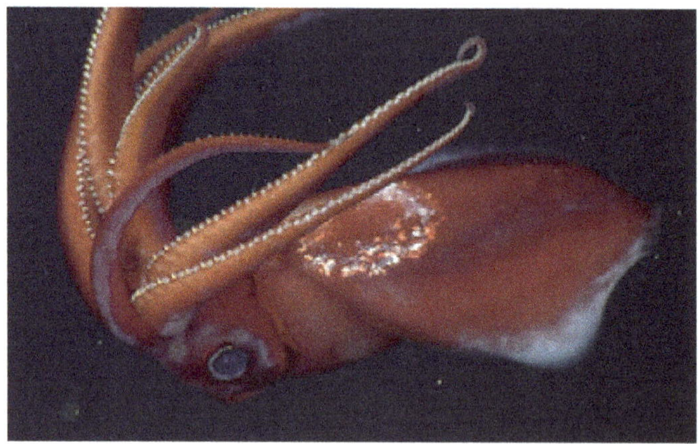

Image credit: MBARI. (Applied for permission)

Common name (s): Not known

Global distribution: Pacific Ocean

Habitat: This mesopelagic species is found at depths of 400 to 800 m

Biology: This species is known to grow to a total length of 24 cm. Male has a penis, which is unusual among squids. A pair of long, elliptical fins runs the length of the body. Tentacles have been reported to drop off in early life stages, and adults possess only eight arms. Long arms bear hooks instead of suckers along most of their length. Body is typically bright red in coloration.

Bioluminescence: This species has numerous photophores, including prominent photophores at the tip of each arm. Further there may be one or two tail photophores (http://dsg.mbari.org/dsg/view/concept/Octopoteuthis%20 deletron).

Ommastrephes bartramii

Image credit: Mary Peart, Wikimedia Commons

Common name (s): Neon flying squid

Global distribution: Circumglobal in subtropical and polar oceanic waters; Pacific, Atlantic and Indian Oceans

Habitat: It occurs at depths ranging from sea level to approximately 2200 m depth. Though it is oceanic, it is known to occur near the surface at night

Biology:. This large pelagic squid is easily distinguishable by the presence of an elongated silver-colored band in the middle of the ventral side of the mantle. Its hectocotylus develops from the left or right fourth arm. Further, the presence of 4 to 7 toothed suckers on the tentacular club is also an important feature of this species. The largest male was found to be 40 cm ML, weighing 2 kg, and the largest female was 61 cm ML, weighing nearly 7 kg.

Bioluminescence: In this species, photophores are present but they are small, irregular, and restricted to the ventral side of the mantle, head, and tentacles. (https://en.wikipedia.org/wiki/Neon_flying_squid). Anon. (https://conxemar.com/en/neon-flying-squid) reported that its luminescent organ is believed to be a long golden or silvery stripe along the ventral midline from mantle opening to level of fin-insertion. Similar golden tissue is also seen on the ventral surfaces of head and ventral arms (IV);. Further, numerous closely-packed, small, very irregularly shaped, often interconnected, light organs are found embedded under the skin in muscle of mantle ventrally. Furthermore, similar light organs occur in patches on ventral surface to head of this species

Rondeletiola minor

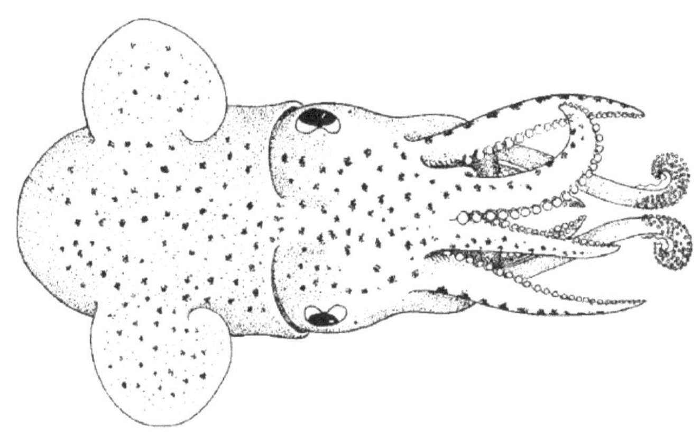

Image credit: FAO. (PD)

Common name (s): Lentil bobtail squid

Global distribution: Eastern Atlantic and the Mediterranean: from Spain to Namibia.

Habitat: This sublittoral, demersal, or upper bathyal species has a depth range of 76 - 496 m. It also lives in brackish-water areas with muddy bottoms.

Biology: Body of this species is short and rounded. Mantle is fused with head dorsally. Fins are kidney-like, short, and are much shorter than the maximum length. Tentacular suckers are in 16 rows. 3rd arm pair in males is usually bent inwards toward mouth.

Bioluminescence: It has a large photophore which is bilobed in juveniles and is compact in adults, with two lenses, in the anterior part of the ink sac (http://species-identification.org/species.php?species_group=zsao and id=2647). Nishiguchi *et al.* (2004) reported that the bacteriogenic light organs of this species are rounded and fused.

Semirossia tenera

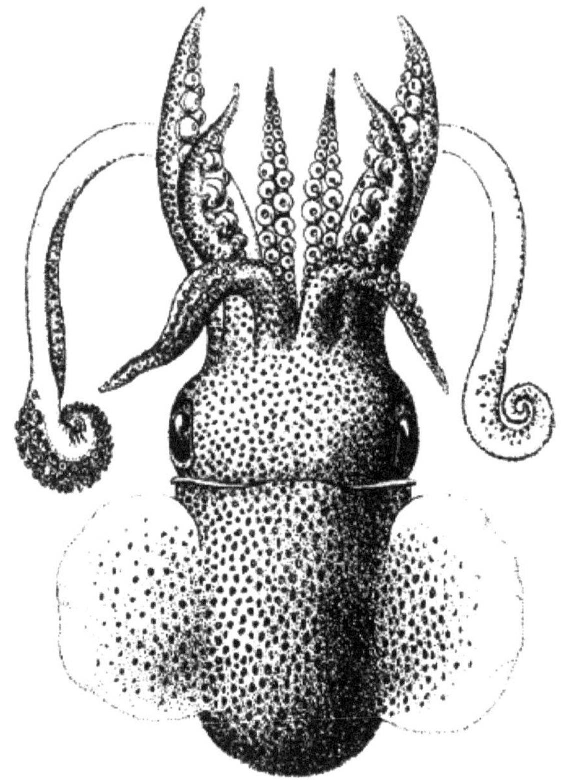

Image credit: FAO (CC)

Common name (s): Lesser bobtail squid or lesser shining bobtail

Global distribution: Arctic, Eastern Central Pacific and Atlantic Ocean: From Nova Scotia to Gulf of Mexico and the Caribbean Sea. Tropical to polar.

Habitat: It inhabits muddy and sandy bottoms on the continental shelf and upper continental slope. It is also believed to bury in soft sediments during the day and emerges at night to forage. This benthic; species has a depth range of 14 - 2622 m

Biology: It grows to 50 mm in mantle length

Oral View

Image credit: Wikimedia Commons

Bioluminescence: This species possesses bacteriogenic light organs. Fusion of the two parts of the light organ across the middle plane was also observed in several specimens of *Semirossia*. (Nishiguchi *et al.*, 2004)

Sepiola ligulata

Common name (s): Not known

Global distribution: Mediterranean Sea.

Habitat: Thei epibenthic species lives on muddy bottoms and mainly on the continental shelf, although occasionally found along the slope. Its depth range is 44 - 380 m

Biology: It is one of the smallest species of this genus. Mantle length of smallest mature males measure 1.10 cm; and smallest mature females are 1.4 cm.

Bioluminescence: This species emits light with its bacteriogenic light organs (Nishiguchi *et al.*, 2004)

Sepiola robusta

Common name (s): Robust bobtail squid

Global distribution: Mediterranean Sea.

Habitat: This benthic species has a depth range of 26 - 498 m

Biology: Females of this species are generally larger than males and the maximum mantle length reported for females was 26.2 mm and 23.5 mm for males

***Sepiola* sp.**

Image credit: Hans Hillewaert, Wikimedia Commons

Bioluminescence: In this species, bacterial light organs are seen which are typically bilobed, kidney-shaped structures embedded in the ink sac. In adult animals, a set of labyrinthine epithelial crypts supports an extracellular culture of luminous bacteria of the genus Vibrio. This epithelial tissue is surrounded by accessory tissues that serve to modify the light. A thick reflector directs the luminescence out of the ventral aspect of the squid; diverticula of the ink sac dynamically rotate around the bacteria-containing epithelium to modulate luminescence intensity; and, a ventral, muscle-derived lens diffuses the light (Foster *et al.*, 2002)

Stauroteuthis syrtensis

Image credit: biographic, California Academy of Sciences. (Applied for permission)

Common name (s): Glowing sucker octopus or bioluminescent octopus

Global distribution: North Atlantic Ocean

Habitat: It is a small pelagic octopus found at great depths

Biology: Mantle length of this species is about 5 to 10 cm and its width about 4 cm. Fins are 4 to 6 cm in width. Its eight arms are of unequal length and the longest one is extending to about 35 cm. Its general texture is gelatinous and the animal is reddish-brown and translucent,

Bioluminescence: This species is known to emit a blue-green light from about 40 modified suckers known as photophores situated in a single row between the pairs of cirri on the underside of each arm. The animal does not emit light continuously, but it can emit for a period of five minutes after suitable stimulation. Some of the photophores emit a continuous stream of faint light, while others are much brighter and switch on and off in a cyclical pattern, producing a twinkling effect. (https://en.wikipedia.org/wiki/Stauroteuthis_syrtensis). In this species, chemical reactions have been reported to trigger an array of blue-green lights that twinkle, pulse, and glow.(https://www.biographic.com/sucker-deception/)

Sthenoteuthis pteropus (=*Ommastrephes pteropus*)

Image credit: B.Christiansen. (Applied for permission)

Common name (s): Orangeback flying squid or orangeback squid,

Global distribution: It is native to tropical parts of the Atlantic Ocean

Habitat: It is found to depths of about 200 m

Biology: Body of this species is bilaterally symmetrical and it has a head with a pair of eyes, eight arms and two tentacles and a fleshy, muscular body known as the mantle. Mantle is cylindrical, narrowing slightly towards the posterior end where there is a wide, roughly diamond-shaped fin. It grows to a mantle length of about 65 cm

Bioluminescence: There are a number of bioluminescent photophores on the head, mantle and fourth arms, with a concentrated patch on the anterior dorsal surface of the mantle forming a luminous orange oval shape (https://en.wikipedia.org/wiki/Sthenoteuthis_pteropus). Roper(1963) reported that it emitted luminous flashes from the ventral surface of the arms and tentacles. Further,when the animals were placed in the water, it ejected several clouds of ink and emitted more flashes of light. The light emitted was pale blue in color. The intensity of the light varied from a faint glow to bright flashes, and the greatest intensity was usually attained whenever the squid was stimulated by touch. The duration of the flashes was two to three seconds, while the dull glow remained in evidence much longer.

Sthenoteuthis oualaniensis (= *Symplectoteuthis oualaniensis*)

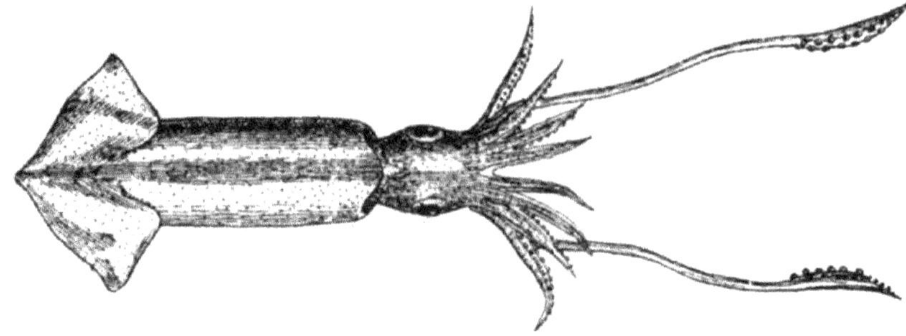

Image credit: Wikiwand

Common name (s): Purpleback flying squid or purpleback squid

Global distribution: Tropical and subtropical Indo-Pacific

Habitat: It lives at depths from the surface to more than 600 m

Biology:. This speices is known to comprise three major and two minor forms with different morphology, anatomy, geographical distribution, and period of spawning. These forms include a giant form, a medium form with a double lateral axis on the gladius, a medium form with a single lateral axis on the gladius, a dwarf form, and a smaller earlier maturing form. Of these the medium form is the typical one in which dorsal mantle length of sexually mature adult males and females is 120 – 150 mm and 190 – 250 mm respectively

Bioluminescence: The medium form is characterized by a dorsal photophore patch; the smaller forms have no dorsal photophore patches; and the giant forms are categorized by a dorsal photophore (https://en.wikipedia.org/wiki/Sthenoteuthis_oualaniensis). Haddock *et al.* (2010) reported that it has a photoprotein that operates with dehydro-coelenterazine which is an oxidised form of coelenterazine. Mirza and Oba (2021) reported that the coelenterazine of this species may undergo an enzymatic oxidation to form dehydrocoelenterazine which is then utilised in its light emission

Stoloteuthis leucoptera

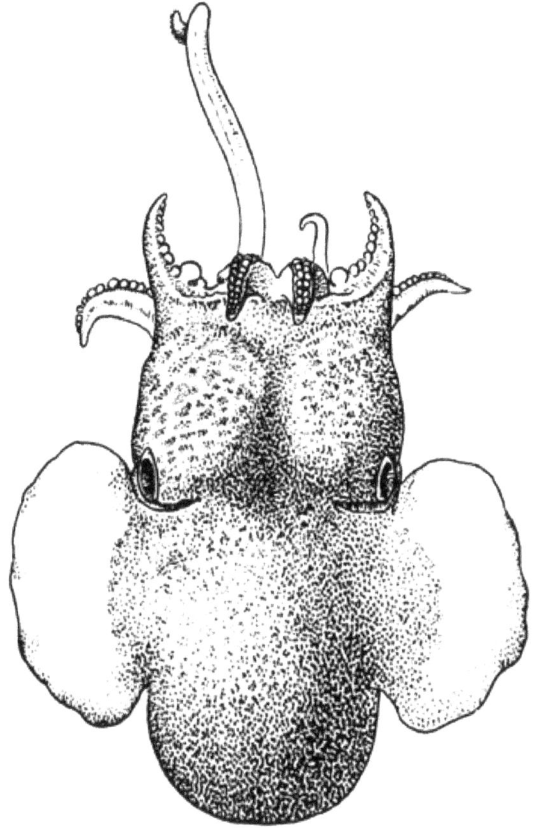

Image credit: Wikiwand

Common name (s): Butterfly bobtail squid

Global distribution: Atlantic Ocean, Mediterranean Sea, and southwestern Indian Ocean.

Habitat: This pelagic species lives in a depth range of 160 - 700 m

Biology: In this species, head is very large and broader than body. Eyes are large and spherical. Fins are large and semicircular. Females are slightly larger than males. They grow to 18 mm and 17 mm in mantle length, respectively.

Bioluminescence: It has two light producing organs under its eyes that contain Vibrio fischeri, a strain of bio-luminescent bacteria (https://animaldiversity.org/accounts/Stoloteuthis_leucoptera/)

Stoloteuthis maoria

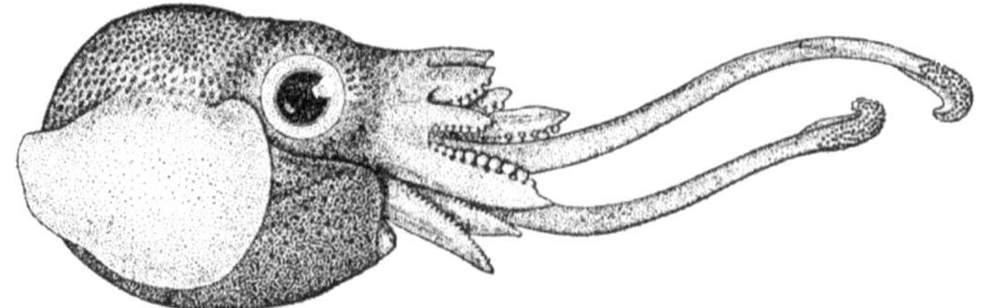

Image credit: R. L. Hudson, Wikimedia Commons

Common name (s): Bobtail squid

Global distribution: It is native to the southwestern Pacific Ocean. It also occurs in New Zealand and eastern Pacific

Habitat: It is commonly found in shallow marine sediments.

Biology: It has very large eyes and protruding fins. Length of the mantle and head–excluding the tentacles–is up to 38 mm, and the width including fins is up to 36 mm. Examination of the type specimens of this species revealed that the type specimens comprised not only two species, but represented two genera. That is the males are the 'real' Stoloteuthis maoria, and the females belonged to an undescribed species, now formally described and named Iridoteuthis merlini.(https://australian.museum/blog/amri-news/fire-shooting-butterfly-bobtail-named-in-honour-of-professor-merlin-crossley/)

Bioluminescence: Like *Iridoteuthis merlini,* this species has been reported to store phosphorescent bacteria in a special organ in its body. It can adjust the luminous intensity of itself, and can eject jets of the luminescent bacteria into the ocean to deter predators (https://en.wikipedia.org/wiki/Iridoteuthis_merlini)

Taningia danae

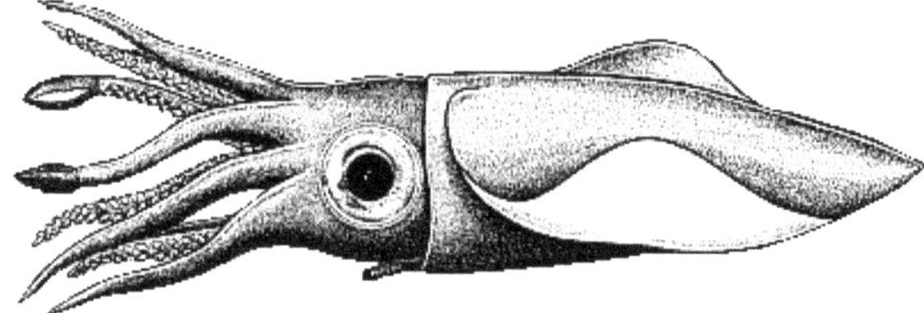

Image credit: Wikimedia Commons

Common name (s): Dana octopus squid

Global distribution: This circumglobal species is mainly found in tropical and subtropical waters including western North Atlantic. It is also found in boreal and notalian waters.

Habitat: This pelagic-oceanic species has a depth range of 385 - 395 m

Biology: It is one of the largest known squid species and it reaches a mantle length of 1.7 m and total length of 2.3 m. It may reach a weight of 61 kg. This highly active squid has claw-like hooks instead of suckers. Not much is known of the biology of Taningia.

Bioluminescence: This species has large (up to 2 cm) yellow, light organs at its arm tips which produced short, bright flashes (Haddock *et al.*, 2010). It is also known to emit long and short intermittent glows; and a steady glow from the visceral photophores (ttps://www.sealifebase.ca/summary/Taningia-danae.htmlh). It is also reported that hese lemon-colored (and lemon-sized) photophores can be flashed at will, because they are equipped with a black, eyelid-like membrane that can be opened or closed (http://www.thecephalopodpage.org/Tdanae.php)

Uroteuthis bartschi

Common name (s): Bartsch's squid

Global distribution: Indo-Pacific.

Habitat: It is a demersal species

Biology: It grows to a known mantle length of 20 cm

Bioluminescence: In this luminescent species, paired,circular ventral photophores are seen (Haneda, 1963; Anderson *et al.*, 2013)

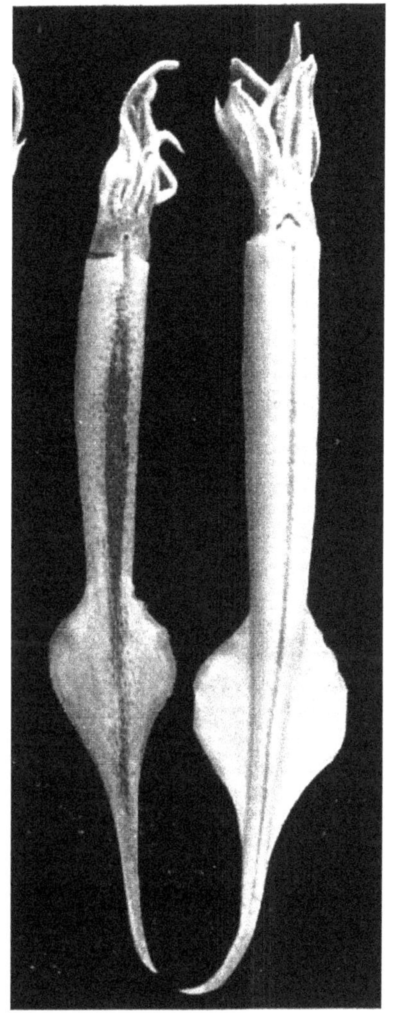

Image credit: Haneda (As the journal belongs to 1963, author is not accessible)

Uroteuthis (Photololigo) duvaucelii (= *Loligo duvaucelii*)

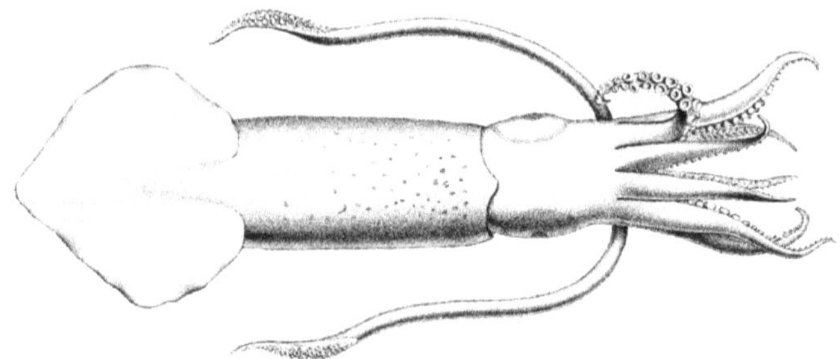

Image credit: A. Pollock, Wikimedia Commons

Common name (s): Indian Ocean Squid or Indian Squid

Global distribution: Indo-West Pacific; throughout the Indian Ocean to Malaysia and the South China Sea; Red Sea and the Arabian Sea.

Habitat: It is found at depths between 3 m and 170 m

Biology: Individuals of this species have mantles which are moderately long (commonly up to 150 mm) and slender with broad fins approximately 50 per cent of the mantle length.

Bioluminescence: In this species, there are two luminous organs which are situated on the ventral surface of the ink sac, near the anus. Each organ consists of a luminous sac which is divided into numerous discrete chambers containing bacteria. On the ventral side,the luminous organ is bordered by a lens-like structure, consisting of muscle cells. On the remaining sides, the bacterial chambers are limited by a cup-shaped reflector layer with numerous parallel lamellae. A reflector separates the bacterial chamber from the ink sac. A ciliated channel connects the interior of the bacterial chamber with the mantle cavity (Pringgenies and Jorgensen, 1994)

Uroteuthis (Photololigo) edulis (= *Loligo edulis*)

Common name (s): Swordtip squid

Global distribution: Western Pacific: Northern Australia, Philippine Islands and northern South China Sea to central Japan.

Habitat: It is a neritic species occurring in 30–170 m depths

Biology: Coloration of the fresh animals is pale reddish brown with numerous brown colored chromatophores scattered all over the dorsal and ventral mantle, head, arms and fins. Mantle is elongated and cylindrical anteriorly and tapering posteriorly to a blunt point. A longitudinal ridge is present along the ventral midline. Fins are large and rhombic. Head is small. Eyes are prominent. Arms are moderately long. Suckers are biserial, oblique and of medium size and are slightly larger in males. Left ventral arm is hectocotylized in males. Tentacles are moderately long and slender; and tentacular clubs are expanded and lanceolate.

Bioluminescence: The luminous bacterial species *Coccobacillus loligo* has been isolated from the luminous gland of this species (Haneda, 1963)

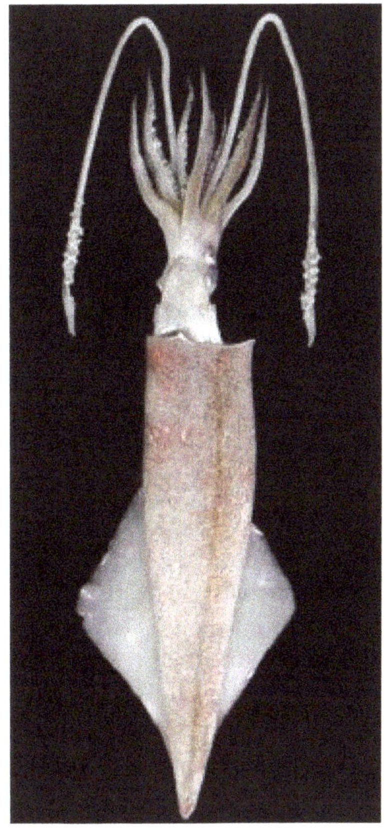

Image credit: Biju Kumar (Not accessible for permission)

Uroteuthis noctiluca (= Loliolus noctiluca)

Image credit: Mark Norman/ Museum Victoria, Wikimedia Commons

Common name (s): Luminous bay squid

Global distribution: It is common in the eastern coast of Australia. Is range is extending from southern Queensland to Tasmania

Habitat: This shallow water species is often seen in seagrass meadows at a maximum depth of about 50 m

Biology: In this species, females are generally grow larger than males. Its maximum mantle length is about 90 mm and width is about 30 mm. Fins are rounded and the posterior of the mantle is heart-shaped. Its eight arms are short and there are rings of suckers on each arm. Each sucker has four to seven teeth. Male has the fourth left arm hectocotylised over its entire length. Its two tentacles are short and robust, with broad paddle-shaped clubs.

Bioluminescence: In this species, the emission of light is due to the presence of luminescent bacteria from its two photophores which are present on the underside of the ink sac. It obtains the luminescent bacteria from the water column (Anderson *et al.*, 2013; https://australian.museum/learn/animals/molluscs/luminous-bay-squid-loliolus-noctiluca-lu-roper-tait-1985/)

Vampyroteuthis infernalis

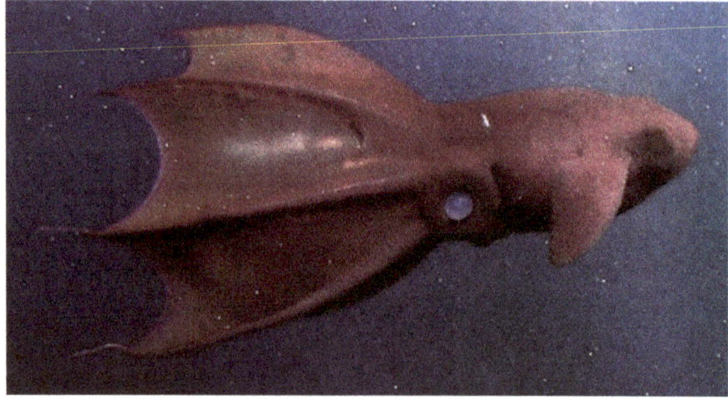

Image credit: Wikiwand

Common name (s): Vampire squid

Global distribution: It is found distributed throughout temperate and tropical oceans

Habitat: It lives in extreme deep sea conditions. It is vertically distributed between the depths of 300-3000m with a majority of individuals however, occupying depths of 1,500-2,500m

Biology: This species can reach a maximum total length around 30 cm. Its 15 cm gelatinous body varies in colour from velvety jet-black to pale reddish. A webbing of skin connects its eight arms and each arm is lined with rows of fleshy spines or cirri.

Bioluminescence: It has photophores *i.e.* large circular organs which are located posterior to each adult fin and are also distributed over the surface of the mantle, funnel, head, and aboral surface. These photoreceptors have been reported to produce luminescent clouds of glowing particles that allow this squid to glow.(https://animaldiversity.org/accounts/Vampyroteuthis_infernalis/). Haddock *et al.* (2010) reported that in this species, there are two large mantle photophores, and small light organs are found scattered across its body. It is known to release glowing particles from special light organs on its arm tip.

Watasenia scintillans

Image credit: Tony Wu. (Applied for permission)

Common name (s): Firefly squid or sparkling enope squid

Global distribution: Japan

Habitat: This deep-sea species spends most of its life in deeper waters between 200 and 400 m.

Biology: Body of this species is bilaterally symmetrical and is divided between a distinct head and a mantle,It has large eyes, eight arms, and two tentacles. Adult firefly squid is approximately 7.5 cm in length and is brown/ red in coloration

Bioluminescence: In this species, its light organs cover the ventral mantle with bright organs near the eyes and at the tips of the arms. It has been reported to produce flashes of blue light via a series of complicated luciferin-luciferase reactions involving ATP, Mg2+, and molecular oxygen (Ding and Liu, 2017; Haddock *et al.*, 2010).

3.8. Echinoderms

Among the echinoderms Asteroidea (starfishes), Ophiuroidea (brittle stars), Crinoidea (sea lilies) and Holothuroidea (sea cucumbers) are bioluminescent. Within these classes, luminous ophiuroids represent more than half of the bioluminescent echinoderms. These luminous echinoderms use mostly coelenterazine dependent bioluminescent systems, although some of them also use photoprotein. Echinoderms inhabiting deep seas are believed to display bioluminescence more commonly. Until recently, information on echinoderm luminescence was entirely descriptive and limited to morphological and ecological observations with few additional remarks on some features of the bioluminescence. This poor documentation of the luminescent phenomenon in echinoderms is largely related to the limited accessibility of species and individuals that can be studied

3.8.1. Asteroids

Class: Asteroidea

Benthopecten sp.

Image credit: NOAA Okeanos Explorer, Wikipedia

Common name (s): Seastar

Global distribution: Common in the coasts of Australia

Habitat: It is a deep-sea dwelling species

Biology: The species of this genus have flexible arms. The inner dorso-lateral surface of the arms possess characteristic longitudinal muscle bands.

Bioluminescence: It is a bioluminescent species (Martini *et al.*, 2019). No other points available

Calyptraster personatus (*Cryptasteria personatus*)

Image credit: RECOLNAT (ANR-11-INBS-0004) - Marie Hennion, Wikimedia Commons

Common name (s): Not known

Global distribution: Western Atlantic: from the Bahamas to Colombia

Habitat: This deep-sea benthos inhabits littoral environments and shallow-water biotopes

Biology: Its depressed form is broadly stellate and abactinal surface is plane. Mouth plates are small and nearly concealed by 3 or 4 pairs of huge, thorny-tipped sub oral spines

Bioluminescence: In this species, a pale blue glow has been observed from its whole dorsal surface when it was disturbed. The brightest areas were however the distal few mm of each arm. Further, in this species, the luminous material was found to appear in its mucus discharge which was mostly seen from the distal region of the arm (Herring, 1974)

Calyptraster sp.

Bioluminescence: In this unidentified species, a feeble bluish glow was observed from the distal region of each arm (Herring,1974)

Diplopteraster multipes

Image credit: Espen Rekdal/Espen Rekdal, Wikimedia Commons

Common name (s): Not designated

Global distribution: Common in Korean Strait

Habitat: It occurs in shallow water as a benthic species

Biology: Body is pentagonal stellate shaped. Each paxilla is composed of a protruded spine, more than eight spiracles, and regularly reticulated muscular bands

Bioluminescence: This species has been reported to produce short bioluminescent flashes in the blue part of the visual spectrum (Birk *et al.*, 2018). Though light could be obtained from both surfaces, ventral surface emitted more light. When the specimen is immersed in freshwater, its ventral surface is known to eject large amounts luminous particles to the exterior in a mucus secretion (Herring,1974)

Freyella elegans (= *Freyella spinosa*)

Common name (s): Starfish

Global distribution: Northwestern Atlantic Ocean

Image credit: NOAA Okeanos Explorer, Wikimedia Commons

Habitat: This deep-sea starfish is living at abyssal depths in the Ocean.

Biology: It has a small disc and 12 long, rigid arms, which are about 210 mm long. Its hexagonal plates on the aboral (upper surface) of the disc bear 1-3 spinelets and are covered by a membrane with no pedicellariae.

Bioluminescence: It is known to emit blue luminescence over the oral and aboral surface of the arms and from the spines (Emson *et al.*, 1995)

Hydrasterias ophiodon

Common name (s): Not designated

Global distribution: North American Atlantic.

Habitat: It is a deep-sea starfish

Biology: Body of the species is orange-colored.

Bioluminescence: In this species, luciferin/luciferase reaction has been observed in its luminescence. The luminescence is visible as a general glow. Further, the luminous material appears on the surface of this species. When it is mechanically stimulated, it emitted a bright blue glow which appeared over the dorsal surface of the disk and the whole length of the arm. (Herring,1974)

Hymenaster koehleri

Common name (s): Sea star

Global distribution: British Columbia

Habitat: It occurs at abyssal depths of about 3000 m

Biology: Not known

Bioluminescence: It is a biololuminescent species (Martini *et al.*, 2019).

Image credit: NOAA (PD), Reproduced with permission

Hymenaster roseus

***Hymenaster* sp.**

Image credit: Adrian James Testa, Wikimedia Commons

Common name (s): Deep-sea slime star

Global distribution: All throughout the world: Atlantic, Pacific, Arctic, Indian and Antarctic

Habitat: It occurs at abyssal depths up to even 8000 m.

Biology: Body of this species is in bright red coloration. Its translucent body largely resembles that of sea cucumbers. The entire surface of this species has evolved into a strange soft covering.

Bioluminescence: When its dorsal surface was stroked, it emitted a faint blue glow in natural conditions. When immersed in freshwater, the whole dorsal surface glowed strongly particularly along the length of each arm. In this species, the luminous material is discharged to the exterior in a mucus secretion (Herring,1974; Emson *et al.*, 1995)

Hymenodiscus coronata (= *Bristingella coronata*)

Image credit: Espen Rekdal/Espen Rekdal, Wikimedia Commons

Common name (s): Not known

Global distribution: Atlantic and the Mediterranean.

Habitat: It is a benthic species with a; depth range of 100 - 2900 m

Biology: Color of the animal is orange to reddish. While the diameter of the disc is 11mm, arms are slender with a length of 9-13 mm.

Bioluminescence: It is known to emit blue luminescence over the oral and aboral surface of the arms and from the spines. Its edge of the disc is also luminous. The eggs dissected from the basal ovaries of this species were also luminescent (Emson *et al.*, 1995)

Novodinia americana

***Novodinia* sp.**

Image credit: Wikiwand

*Common name (s):*Asteroid

Global distribution: Arctic sea, Greenland.

Habitat: Deep-water starfish species inhabiting the aphotic zone,

Biology: It has well developed eyes at the end of its arms and these eyes are composed of small units (ommatidier) that produce the equivalent of a high definition picture. Its eyes supports sight that is better or just as good as starfish living in shallow sunlit waters.

Bioluminescence: It is highly bioluminescent with more or less the entire body surface capable of emitting light for an extended period of time (several seconds). It was also in the blue (short wavelength) part of the spectrum. The emission in this species is having a longer lasting glow (several seconds), (Birk *et al.*, 2018; https://sciencenordic.com/animals–plants-denmark-greenland-science-special/stunning-starfish-illuminates-the-dark-arctic/1454536)

Pectinaster filholi (= *Pectinaster forcipatus*)

Common name (s): Asteroid

Global distribution: North-East Atlantic Ocean

Habitat: It is a deep-sea asteroid

Biology: Not reported

Image credit: RECOLNAT (ANR-11-INBS-0004) - Marie HENNION, Wikimedia Commons

Bioluminescence: In this species, the luminescence is visible as a general glow. Further, its luminescence was found restricted to very limited areas of the body and is associated with particular surface structures. When its dorsal surface of the disc is mechanically stimulated, a spot of blue-green light appeared at the base of the arms(Herring,1974)

Plutonaster agassizi (= *Plutonaster notatus*)

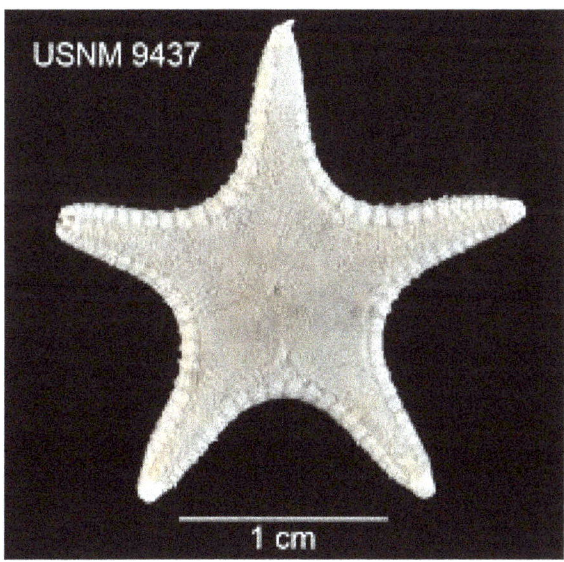

Image credit: EOL. (CC)

*Common name (s):*Asteroid

Global distribution: Mexico and Argentina; North Atlantic Ocean

Habitat: This benthic species occurs even at a depth of 3300 m.

Biology: Body has pentaradial symmetry.

Bioluminescence: By firm stroking of the dorsal surface, this species emitted blue light and this light was in greater density when the specimen was immersed in freshwater. Further, the light appeared as a steady glow over the entire dorsal surface. Light was also visible from the arms in dorsal, ventral and lateral views. In this species, the control of luminescence has been reported to be neuronal (Herring,1974)

Plutonaster bifrons

Image credit: Wikidata

Common name (s): Asteroid

Global distribution: Atlantic and the Mediterranean Sea.

Habitat: This benthic species has a depth range of 10 - 500 m

Biology: It is essentially a scavenger of organic remains, including fish and feeds at or close to the sediment surface

Bioluminescence: This luminescent species has a λmax of 525nm (Johnsen *et al.*, 2012)

Zoroaster fulgens

Common name (s): Asteroid

Global distribution: Western Atlantic and Antarctic Atlantic: Gulf of Mexico and Caribbean.

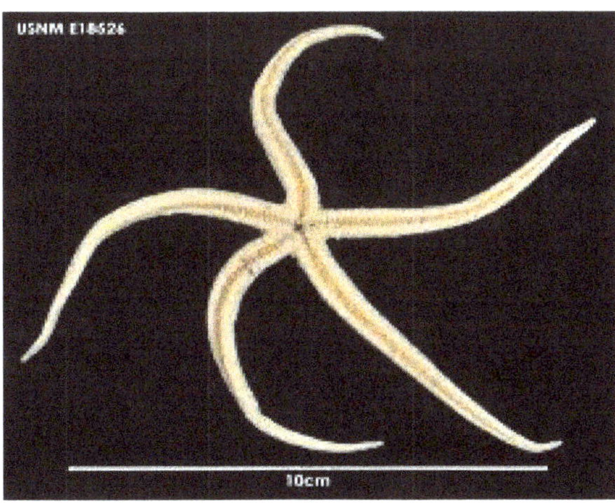

Image credit: EOL. (CC)

Habitat: This benthic; species has a depth range of 220 - 3000 m. It is found in association with shallow-water mollusks, decapod crustaceans and colonial corals

Biology: Specimens of this species have a size range of 12 mm -114 mm.

Bioluminescence: It emits light as bright points over the general aboral surface (Emson *et al.*, 1995)

3.8.2. Ophiuroids

Class: Ophiuroidea

Acrocnida brachiata

Image credit: Picton, B.E. and Morrow, C.C. National Museum of Northern Island (Applied for permission)

Common name (s): Sand burrowing brittlestar

Global distribution: Northeastern Atlantic Ocean and the North Sea

Habitat: It occurs on the seabed and is living semi-buried in the sand with only its arm tips projecting. Its depth range is from the lower shore down to about 40 m

Biology: This greyish-brown colored species has a flat disc up to 12 mm in diameter and five slender, clearly-demarcated, articulated arms up to 180 mm in length. The arms of this species which flex sideways rather than up and down, have a pair of tentacle scales on each joint as well as numerous spines.

Bioluminescence: In this speices, the luminescence is confined to its arm spines (Emson and Herring,1985),

Amphipholis squamata

Image credit: Blair Patullo, Wikimedia Commons

Common name (s): Brooding snake star or dwarf brittle star,

Global distribution: British Isles and Ireland.

Habitat: It lives in the intertidal zone in shallow water, and can be seen under large stones, shells, and around bryozoans.

Biology: Disk of this species is round and is covered by coarse scales. Disc's diameter is up to 5 mm and is usually smaller. Radial shields extend about one third to half the radius of the disk. They are usually slightly longer than broad. The ventral interradial plates are slightly smaller than the dorsal plates and there is usually a distinct boundary between them.

Bioluminescence: In this species, luminescence occurred only in response to mechanical stimulus and it ranged from weak to intense reaction depending on the contact organism. Interestingly, only crustaceans have been reported to trigger an intense luminescence. Further, brachial autotomy also occurred after prolonged interaction and the luminescence continued only from the autotomized arm.(Deheyn *et al.*, 2000)

Brehm and Morin (1977) reported that the luminescence of this species is intracellular and is largely due to photogenic cells, termed photocytes. Further, the emission spectra of luminescence in this species is broad with a half band width of 71 nm and an approximate emission maximum at 510 nm. Furthermore, the fluorescence of this species appeared only after the onset of luminescence.

Amphiura arcystata

Image credit: Stanford, SeaNet. (Applied for permission)

Common name (s): Not known

Global distribution: NE Pacific: California, Gulf of California to Colombia, Cocos and Ecuador (Galapagos Islands)

Habitat: It occurs at depths of 6-849m. It is very abundant in sand where water motion is low

Biology: It has very long arms which grow up to 30 cm and are protruding from sand.

Bioluminescence: It is a bioluminescent species (Tsuji *et al.*, 2005)

Amphiura constrica

Common name (s): Not known

Global distribution: Around Australia.

Image credit: Blair Patullo, Wikimedia Commons

Habitat: It is found distributed in clumps of algae, seagrass and encrusting invertebrates or under stones from intertidal rock pools to the inner continental shelf (0-60 m)

Biology: It is a small species with a disc covered in numerous tiny plates. There are no spines. Arms are moderately long and sinuous. There are up to 8 arm spines on the side of each segment, Colour of the animal is pale grey or brown disc with dusky banded arms. Disc is up to 7 mm wide, arm up to 5 cm long.

Bioluminescence: It is a bioluminescent species (Tsuji *et al.*, 2005)

Amphiura filiformis

Image credit: iNaturalist (CC)

Common name (s): Brittle star

Global distribution: North east Atlantic Ocean

Habitat: It is found on the seabed up to a depth of 200 m. It is known to dig itself a shallow burrow (5 cm) in the sand and waves its arms in the water above

Biology: Disk of this species is round to pentagonal, and its diameter is up to 10mm. Centrodorsal and primary plates are conspicuous. Radial shields are just over twice as long as broad and are separated except at their distal ends. Ventral interradial areas are partially naked, but in some larger specimens, they are often covered with overlapping plates.

Bioluminescence: It emits a bluish light (https://en.wikipedia.org/wiki/Amphiura_filiformis). In this species, its luminescence is confined to the arm spines. Further, the light emission appears to be a function of basophilic gland cells in the spines which are known to contain spherical granular bodies. (Emson and Herring,1985). Delroisse *et al.* (2017) reported that in this species, luciferase is expressed at the base of the spine and at the arm tips within well-defined photocyte clusters. Mallefet *et al.* (2020) reported that it is known to use coelenterazine as substrate of a luciferin/luciferase luminous system.

Amphiura (Amphiura) grandisquama

Image credit: RECOLNAT (ANR-11-INBS-0004) - Marie HENNION, Wikimedia Commons

Common name (s): Not designated

Global distribution: Atlantic, Indian and Pacific Oceans

Habitat: It is a deep-sea species with a depth range of 300-1600 m. It is also found associated with carbonate chimneys and crusts

Biology: Disk of this species is round, indented over the arms, and is covered with small scales. Disk's diameter is up to 7mm. Centrodorsal and primary plates are usually indistinct. Radial shields are about 3 times as long as broad and are fully separated. They extend for about half the radius of the disk. Ventral interradial areas are covered by small scales.

Bioluminescence: It is a bioluminescent species (Emson and Herring, 1985).

Amphiura josephinae

It is a bioluminescent species (Tsuji *et al.*, 2005). No other information is available

Amphiura kandai

It is a bioluminescent species (Tsuji *et al.*, 2005). No other information is available

Amphiura magellanica

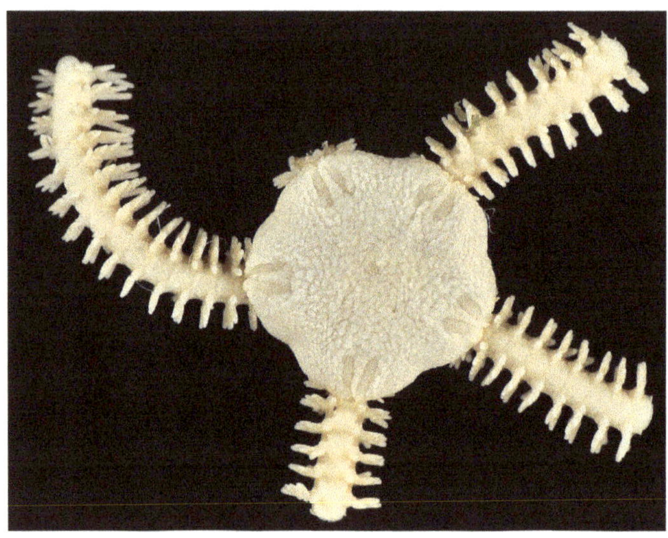

Image credit: Adrian James Testa, Wikimedia Commons

Common name (s): Not known

Global distribution: New Zealand, south-eastern Australia and South America

Habitat: This benthic species has a depth range of 1- 1400 m.

Biology: Its disc is finely scaled dorsally and ventrally. There are 6–7 arm spines, reducing to five by mid-arm. Ventral most arm spine is elongated and projecting downward on mid arm segments. There is a single, large, leaf-like tentacle scale.

Bioluminescence: It is a bioluminescent species (Tsuji *et al.*, 2005)

Ophiarachnella ramesayi

It is a bioluminescent species (Tsuji *et al.*, 2005). No other information is available

Ophiacantha abyssicola

Common name (s): Not reported

Global distribution: Both sides of the North Atlantic

Habitat: It is a predominantly bathyal species occurring at a depth range of 260-2300 m

Biology: Disk of this species is round and is up to 8 mm in diameter. It is covered with glassy spinelets, Plates of the interradial region are usually large and conspicuous.

Bioluminescence: It is a bioluminescent species (Tsuji *et al.*, 2005)

Ophiacantha aculeata

Common name (s): Not designated

Global distribution: West Atlantic from off Virginia

Habitat: It has been reported from depths of 1650- 3600 m.

Biology: Disk of this species is round to slightly pentagonal, with a diameter up to 13mm. It is covered by trifid spinelets. Only the tips of the radial shields are visible. Ventral interradial areas are also covered with spinelets.

Bioluminescence: This species is known to emit a bright blue-green light. Individual specimens flashed brightly initially and then emitted a steady but feeble blue-green glow. The light appeared at the base of the spines on the dorsal surface of the segment. Further the light was visible from both the aboral and oral sides. In this species, the control of luminescence has been reported to be neuronal. Further, in this species a luminous fluid is found secreted on treatment with freshwater (Herring,1974; Tsuji *et al.*, 2005)

Ophiacantha aristata

*Common name (s):*Not known

Global distribution: North-east Atlantic

Habitat: It is known to occur at a depth range of 822-1700 m.

Biology: Disk of this species is round, and is covered with spinelets. Disk diameter is up to 8 mm. Spinelets are crowned with 9 or more points. Radial shields are sometimes inconspicuous but in some specimens their distal ends are distinguishable. Ventral interradial areas are covered with spinelets which are similar to those of the dorsal side.

Bioluminescence: It is a bioluminescent species (Tsuji *et al.*, 2005)

Ophiacantha bairdi

It is a bioluminescent species (Tsuji *et al.*, 2005). No other information is available

Ophiacantha bidentata (= *Ophiacantha fraterna*)

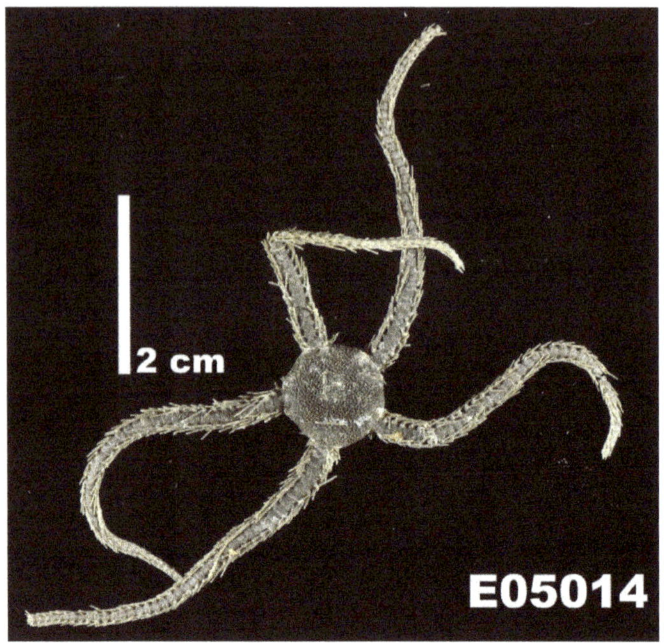

Image credit: Georgia Spence, Wikimedia Commons

Common name (s): Not designated

Global distribution: It is a widespread arctic-boreal ophiuroid with a circumpolar distribution; Arctic seas, North Atlantic and North Pacific.

Habitat: Dense populations of this species are known to occur in the hundreds to thousands of individuals in the coral habitats of shallow and deeper waters. Its depth range is 32-4730m.

Biology: Disk of this species is round, and is covered with an armament of spinelets. Disk diameter is up to 13 mm. Spinelets are trifid but they may be elaborated with secondary points. Radial shields are long, thin and separated. Ventral interradial areas have similar armament to the dorsal side.

Bioluminescence: When this species was placed in freshwater, a steady blue glow in the form of two broad longitudinal streaks at the lateral margins of each segment was observed. Further light was also observed from the spines, basal plates and lateral arm plates. Further more light has also been observed from the podia of this species (Herrng,1974; Tsuji *et al.*, 2005)

Ophiacantha crassidens

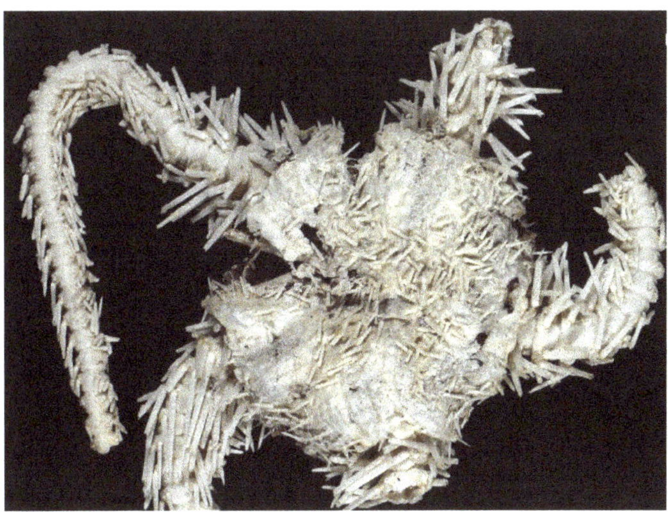

Image credit: Av RECOLNAT (ANR-11-INBS-0004) - Marie Hennion, Wikimedia Commons

Common name (s): Not known

Global distribution: Both sides of the North Atlantic

Habitat: This benthic species has a depth range of 980-3120m.

Biology: Disk of this species is round and its diameter is up to 14mm. Disc is covered by small imbricating scales. Only the tips of the radial shields are visible. Ventral interradial areas are covered by scales and spinelets which aresimilar to those of the dorsal surface.

Bioluminescence: It is a bioluminescent species (Tsuji *et al.*, 2005)

Ophiacantha cuspidata

*Common name (s):*Not known

Global distribution: Eastern Atlantic

Habitat: This benthic species occurs at a depth range of 768-2460 m.

Biology: Disk of this species is round and its diameter is up to 10mm. It is covered by large rugose spinelets which are with elaborate multipointed crowns. Tips of the radial shields are sometimes visible. Ventral interradial areas are covered by spinelets similar to the dorsal side

Bioluminescence: It is a bioluminescent species (Tsuji *et al.*, 2005)

Ophiacantha densa

Common name (s): Not known

Global distribution: Eastern Atlantic

Habitat: This benthic species occurs at a depth range of 300-1 324 m.

Biology: Disk of this species is round and is diameter is up to 5-6 mm. This disk is covered with a dense coating of spinelets. Radial shields are inconspicuous. Ventral interradial areas are also covered with spinelets similar to those of the dorsal surface.

Bioluminescence: It is a bioluminescent species (Tsuji *et al.*, 2005)

Ophiacantha simulans

Common name (s): Not known

Global distribution: Both sides of the Atlantic

Habitat: This benthic species is known to occur at a depth range of 1575-3018 m.

Biology: Disk of this species is round and its diameter is up to 10 mm. It is covered with a dense coating of spinelets which are crowned with 7-8 points. Radial shields are long and narrow with only their distal ends visible. Only the distal part of each ventral interradial area is covered with spinelets.

Bioluminescence: It is a bioluminescent species (Tsuji *et al.*, 2005)

Ophiacantha smitti

Common name (s): Not recorded

Global distribution: Both sides of the North Atlantic

Habitat: It is having a depth range of 994-2282 m

Biology: Disk of this species is round and its diameter is up to 8 mm. Disk is covered by a dense coating of simple rods and trifid and bifid spinelets. Spinelets are normally situated periferally whereas the rods are found in the middle of the disk. Both rods and spinelets may be covered with a coating of skin

Bioluminescence: It is a bioluminescent species (Tsuji *et al.*, 2005)

Ophiochiton ternispinus

Common name (s): Not reported

Global distribution: North Atlantic

Habitat: This benthic species has a depth range of 425-2220m

Biology: Disk of this species is round and flat, and its diameter is up to 20 mm. Disk is covered with many small imbricating plates in large specimens or by fewer larger plates in smaller ones. Centrodorsal and primary plates are distinct and larger than the surrounding plates. Radial shields are small and are well separated from one another; and they are usually longer than broad, and slightly teardrop shaped.

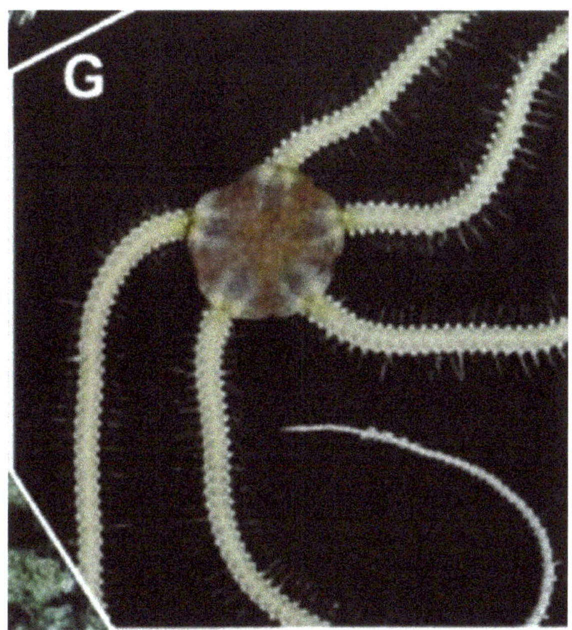

Image credit: Sonke Johnsen, Reproduced with permission

Bioluminescence: It has been reported to emit blue light and its emission peak wavelength was 470 nm (Johnsen *et al.*, 2012)

Ophiocomina nigra

Common name (s): Black brittle star or black serpent star

Global distribution: It occurs in the north-eastern Atlantic Ocean, the North Sea and the Mediterranean Sea; it is also common in the English Channel

Habitat: It is found on rocks, boulders and gravel in the shallow neritic zone down to about 100 m but it may occasionally be found at greater depths. It is believed to be tolerant of low salinity levels.

Biology: It is a large brittle star with five narrow arms which are up to 125mm long. Its distinct central disc is up to 25mm wide. The general colouration of this species is black or with shades of brown. Pale coloured specimens also occasionally occur. Upper surface of the disc is covered with fine granules. On the ventral side, the granules are found restricted to the outer portion and the plates are visible towards the central mouth.] There is a comb-like arrangement of spines down either side of the arms and this gives them a bristly appearance. On the upper side, each arm segment is covered by a broad plate with 5 to 7 spines. On the underside there are tube feet which are without suckers

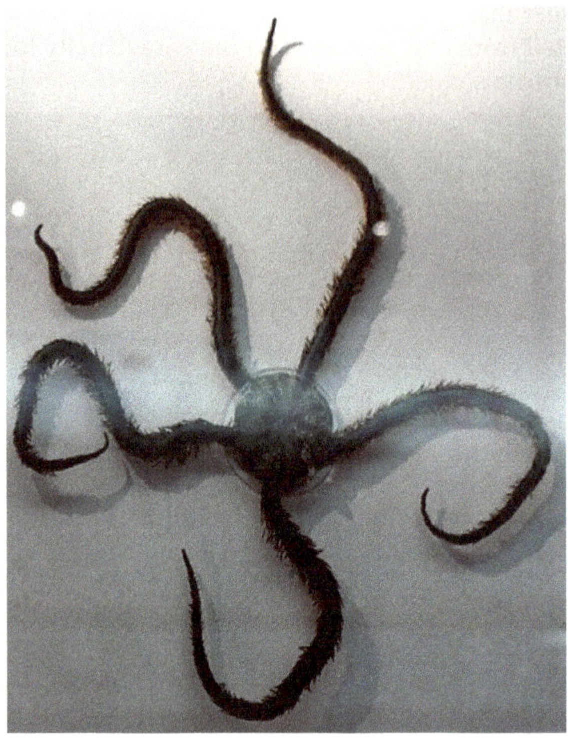

Image credit: Emõke Dénes, Wikimedia Commons

Bioluminescence: It is known to emit light through chemical triggering by hydrogen peroxide. The luminous capabilities of this species are homogeneously spread along its arms. The mechanical stimulation of arms before chemical triggering has been reported to strongly enhance the luminous capabilities of this species. Further, its luminous mucus emission is said to be associated with other defensive function (Jones and Mallefet, 2012)

Ophiomitra spinea

Common name (s): Not reported

Global distribution: Both sides of the Atlantic

Habitat: It occurs at a depth range of 2500-3150m

Biology: Disk of this species is roun and is indented interradially. Area of the radial shields is raised up. Diameter of the disk is up to 5 mm. Disk is covered by small overlapping plates. Rradial shields are naked. They are large nearly as wide as long. Ventral interradial areas are covered with small plates and spinelets similar to those of the dorsal side.

Bioluminescence: It is a bioluminescent species (Tsuji *et al.*, 2005)

Ophiomitrella sp.

It is a bioluminescence species (Tsuji *et al.*, 2005). No other information is available

Ophiomusa lymani (= *Ophiomusium lymani*)

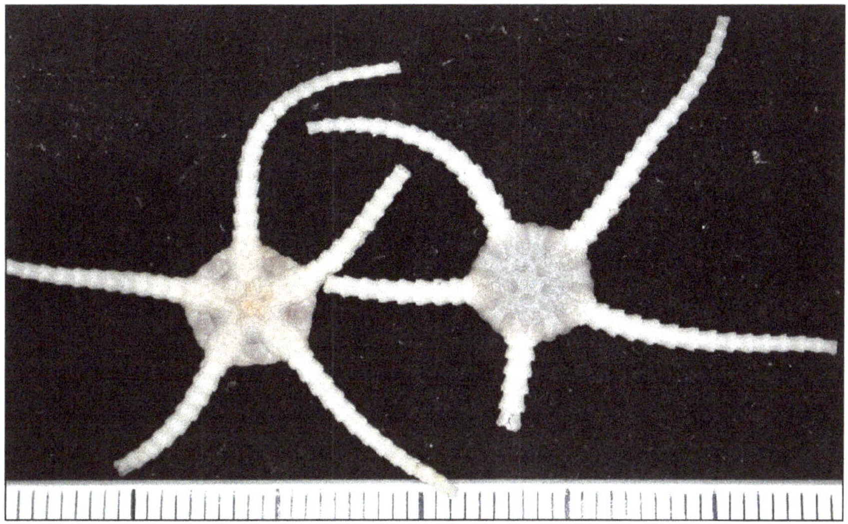

Image credit: CSIRO - Karen Gowlett-Holmes, Wikimedia Commons

Common name (s): Not known

Global distribution: Atlantic, Pacific and Indian Oceans.

Habitat: This benthic species has a depth range of 130-4800 m

Biology: Disc diameter of this species ranges from 17 to 20 mm. The arm spine articulations are placed at the distal edge of the lateral arm plate. Disk of this species is pentagonal, and is high covered with relatively small plates. Its diameter is up to 30 mm. Radial shields are large, longer than broad and are slightly triangular in shape. All the dorsal disk plates are texture. There is a rudimentary arm comb of small, rounded, irregularly arranged papillae.

Bioluminescence: It is a bioluminescent species (Tsuji *et al.*, 2005)

Ophionereis fasciata

Common name (s): Mottled brittle star

Global distribution: Southwest Pacific: New Zealand and Kermadec

Habitat: This benthic species occurs on open shores down to a depth of about 300 m

Biology: Disk of this species has pentaradial symmetry

Image credit: Auckland War Memorial Museum, Wikimedia Commons

Bioluminescence: It is a bioluminescent species (Shimomura and Yampolsky, 2019)

Ophionereis schayeri

Image credit: Museums Victoria. (CC)

Common name (s): Banded Brittle Star

Global distribution: Southern Australian coasts.

Habitat: It occurs on rocky areas, often under flat rocks, up to a depth of 180 m.

Biology: Disc of this species is covered with minute scales. Arms are with 4 small blunt arm spines on each side. The plates along the arms are in three pieces. Body of this species is grey to cream with wide dark stripes around arms. While disc is up to 2.5 cm wide, arms are up to 15 cm long.

Bioluminescence: This species has been reported to drop off the ends of their arms, like lizards drop off their tails. These arm pieces possess a green luminescent glow (https://collections.museumsvictoria.com.au/species/8642; Tsuji *et al.*, 2005)

Ophiopholis longispina

It is a bioluminescent species (Tsuji *et al.*, 2005). No other inormaionn is available

Ophioplinthaca chelys

It is a bioluminescent species (Tsuji *et al.*, 2005). No other inormaionn is available

Ophioplinthaca rudis

Image credit: RECOLNAT (ANR-11-INBS-0004) - Marie Hennion, Wikimedia Commons

Common name (s): Not known

Global distribution: Western Central Pacific: Philippines and New Caledonia; South China Sea.

Habitat: This benthic and deep- water species has a depth range of 650 - 3124 m

Biology: Its disc spines are long, slender, needle-like, and are smooth to finely serrate. Its tentacle scale is bottle-shaped to pointed

Bioluminescence: The bioluminescence of this species has been reported to be under cholinergic control (Pablo, 2020)

Ophioplinthus tessellata (= *Homalophiura tesselata*)

Image credit: Eric A. Lazo-Wasem, Wikimedia Commons

*Common name (s):*Not known

Global distribution: North Atlantic,

Habitat: It occurs in soft substrates at a depth range of 433– 4,706 m

Biology: It has strongly reduced arm comb and tentacle pores are restricted only to the few proximal segments. Disk of this species is pentagonal. Sometimes it is high, and covered with large irregularly shaped plates. Disk diameter is up to 30 mm. Radial shields are small. There are a number of block-like papillae which are visible opposite the base of each arm forming a rudimentary arm comb. These may be arranged into two rows.

Bioluminescence: When this species was treated with freshwater, quantities of luminous particles stream out from beneath the supradorsal membrane while the arm glows steadily as two lines each side of the central ossicles and brightest at the distal tip (Emson *et al.*, 1995; Tsuji *et al.*, 2005)

Ophioplocus bispinosus (= *Ophioceres bispinosus*)

Common name (s): Not reported

Global distribution: South-eastern Australia.

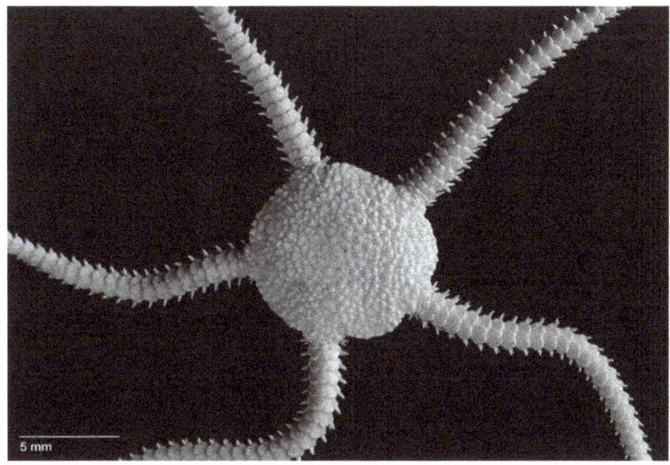

Image credit: Museums Victoria (CC)

Habitat: It occurs under rocks on reefs to a depth of about 50 m.

Biology: Disc of this species is round and flat. Arms are short and stiff.; and slow moving and relatively rigid. Disc is covered in minute tumid plates. Arm plates are fragmented into numerous platelets. There are two short arm spines on each side of a segment. Colour of this species is uniform grey, tan or pale; and arms may be faintly banded. While disc is up to 1 cm in diameter, arms are up to 2 cm long.

Bioluminescence: It is a bioluminescent species (Tsuji *et al.*, 2005)

Ophiopsila annulosa

Image credit: Bernard Picton, Reproduced with permission

Common name (s): Not known

Global distribution: It occurs in the Mediterranean Sea and adjacent Atlantic Ocean north to western Scotland

Habitat: This species has been reported to have strict habitat requirements as it is living only in coarse gravel. It is known to quickly retract into the gravel on disturbance.

Biology: It is a large brittle star with long banded arms which are with light and dark brown and there is a reticulate mottled pattern of brown on the disc. Arm spines are flattened and arranged in groups of 11-12. Tentacle scales are longer and are exceptionally large, supporting the tube feet when the animal is suspension feeding. Central body is 14mm in diameter with arms which are approximately 140 mm in length

Bioluminescence: Spine luminescence has been observed in this species (Gotto,1963; Tsuji *et al.,* 2005; Herring,1974)

Ophiopsila aranea

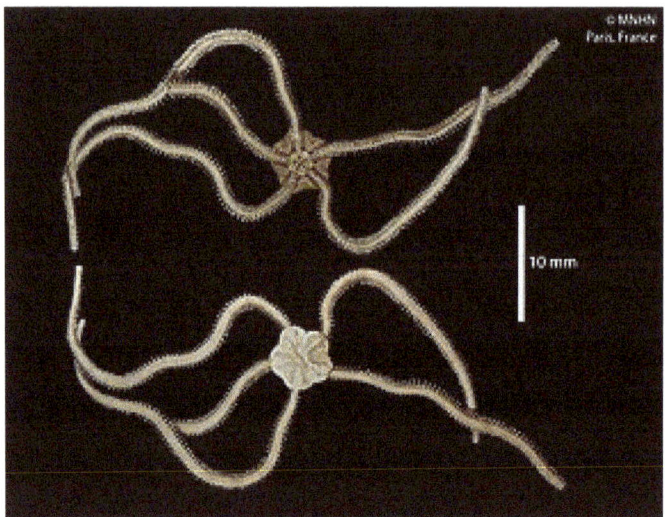

***Ophiopsila* sp.**

Image credit: Wikipedia

Common name (s): Not known

Global distribution: Eastern Mediterranean and Atlantic coasts

Habitat: It lives in crevices extending its long arms out into the water. It has a depth range of 9 - 185 m

Biology: It has a round disc, Arms are banded with light and dark brown and there is a reticulate mottled pattern of brown on the disc. Arm spines are flattened and arranged in groups of 6-8. Tentacle scales are shorter and are exceptionally large. While the disc has a diameter of 10mm. arms are about 90 mm long

Bioluminescence: In this species, KCl application induced luminescence in isolated arms. The light emission was found characterized by a series of flashes whose maximal intensity increased as a function of KCl concentration (Mallefet and Isson,1995). Jones and Mallefet, (2010) reported aposematic use of bioluminescence in this species. This species emitted light only after disturbance, and remained cryptic the rest of the time.

Ophiopsila californica

Common name (s): Not known

Global distribution: Eastern Central Pacific: USA.

Habitat: This benthic species is found buried in soft bottoms, sand or mud, with only arms exposed, in shallow subtidal depths

Biology: Not reported

Bioluminescence: Brehm and Morin (1977) reported that the luminescence of this species is intracellular and is largely due to photogenic cells, termed photocytes. Further, the emission spectra of luminescence in this species is broad with a half band width of 71 nm and an approximate emission maximum at 510 nm. Furthermore, the fluorescence of this species appeared only after the onset of luminescence.

Shimomura (2008) reported that this species when treated with H2O2 emitted green light with a λmax of 500 nm. which indicated that the green fluorescent chromophore is the light-emitter of the photoprotein luminescence

Emson(1995) reported that this species showed bright luminescence in contact with polychaetes.

Ophiopsila riisei

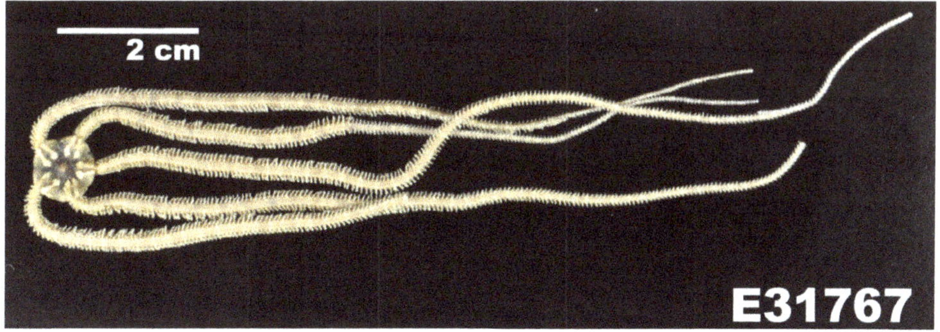

Image credit: EOL (CC)

Common name (s): Not reported

Global distribution: Western Central Atlantic: Belize.

Habitat: it is a reef-associated species with a depth range of 1 - 366 m. This species is typically found in rocky substrates in rubble and is often associated with soft, peat bank substrates. It is also common in the clumps of seaweeds

Biology: Not reported

Bioluminescence: It is a bioluminescent species (Tsuji *et al.*, 2005). Grober (1988) reported that the luminescent signals produced by this species functioned as an aposematic deterrent against crustacean predators.

Ophiopsila xmasilluminans

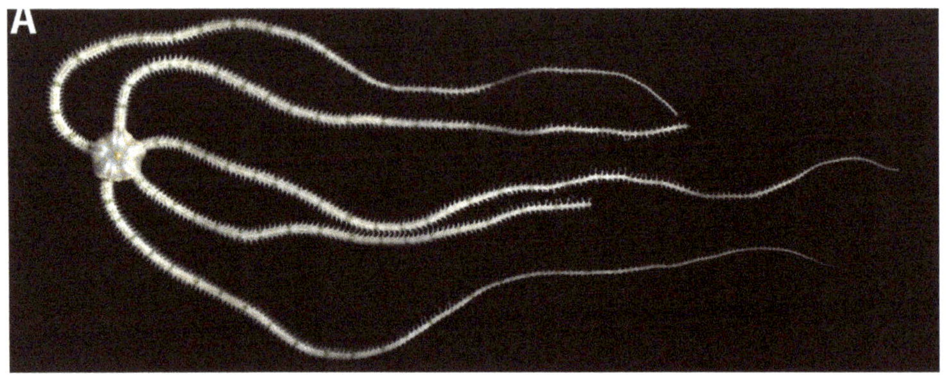

Image credit: Mokanishi (Applied for permission)

Common name (s): Not reported

Global distribution: Christmas Island in northwestern Australia

Habitat: It occurs on sandy bottoms in submarine cave with its disc buried and arms extended above the substratum.

Biology: This species possesses a disc which is entirely covered by thick skin with small and delicate embedded scales and granules. There are 3 oral papillae which are basically flat, fan-shaped. Arm spine is long and flat. Inner tentacle scale is narrow, flat and long. Oral disc is creamy white and yellowish spot is seen on oral interradial disc. Arms are 85 mm long and disc diameter is about 5mm

Bioluminescence: It is a bioluminescent species (Okanishi *et al.*, 2019)

Ophioscolex glacialis

Common name (s): Not designated

Global distribution: North Atlantic and from Arctic Seas

Habitat: This benthic species occurs at a depth range of 50 to about 2,727 m

Image credit: Sorcha Cronin-O'Reilly, Reproduced with permission

Biology: Disk of this species is round to pentagonal, delicate and is covered with skin without any plates. Disk diameter is up to 25 mm. Radial shields are inconspicuous. Ventral interradial areas are also covered by naked skin.

Bioluminescence: Spine luminescence has been observed in this species (Gotto, 1963; Tsuji *et al.*, 2005; Herring,1974)

Ophiothrix fragilis

Image credit: Lamiot, Wikimedia Commons

Common name (s): Common brittle star or hairy brittle str

Global distribution: Around the coasts of western Europe; along the coast of South Africa

Habitat: It is most common on rock and on coarse sediments. It is often found in empty shells or under stones, from the littoral zone down to a depth of about 350 metres

Biology: It is extremely variable in colouration, ranging from violet, purple or red to yellowish or pale grey,; and is often spotted with red. Arms are usually white or grey with pink bands. Central disc is about 1 cm in diameter and its five arms are about 5 cm long. Its slender tapering arms are quite distinct from the disc and are covered with overlapping scales.Arms are extremely fragile and are easily shed, coming away either whole or in pieces.[

Bioluminescence: It is a bioluminescent species (Tsuji *et al.*, 2005)

Ophiura (Ophiura) mundata

Image credit: RECOLNAT (ANR-11-INBS-0004) - Marie Hennion, Wikimedia Commmons

Common name (s): Not reported

Global distribution: It has been recorded from the Labrador Basin, SW of Iceland, from the Bay of Biscay, off Portugal south to the Azores

Habitat: This benthic species occurs at a depth range of 2043-4315 m.

Biology: Disk of this species is round to subpentagonal and is covered with large plates. Disk diameter is up to 8 mm. Radial shields are longer than broad. There is a single row of block-like contiguous arm comb spinelets opposite each side of the arms.

Bioluminescence: In this species, quantities of luminous particles stream out from beneath the supradorsal membrane when it is treated with freshwater. The arm glows steadily as two lines each side of the central ossicles (Emson *et al.*, 1995; Tsuji *et al.*, 2005).

Ophiuroglypha irrorata concreta (= *Ophiura concreta*)

Common name (s): Not known

Global distribution: It has cosmopolitan distribution in all oceans except Arctic.

Habitat: It occurs in soft substrates at a depth range of 403– 7,340 m. However, it is most common at depths over 2,000 m

Biology: Disk of this species is pentagonal and is covered with relatively large plates. Disk diameter is up to 20 mm. Radial shields are small, teardrop shaped and are longer than broad. Arm comb consists of a single row of contiguous block-like spinelets.

Bioluminescence: In this species, fairly larger quantities of luminous particles stream out from beneath the supradorsal membrane when it is treated with freshwater. Subsequently the arm glows steadily as two lines each side of the central ossicles (Emson *et al.*, 1995; Tsuji *et al.*, 2005)

3.8.3. Crinoids

Class: Crinoidea

Endoxocrinus (Diplocrinus) wyvillethomsoni (= *Annacrinus wyvillethomsoni*)

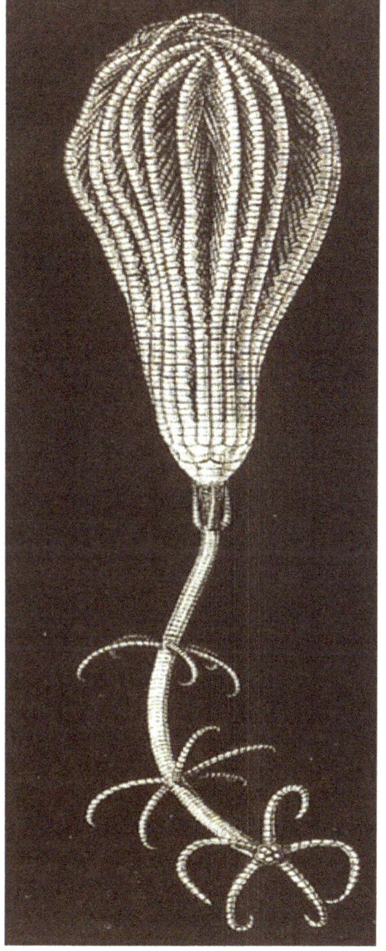

Image credit: F. Huth, Wikimedia Commons

Common name (s): Blue-green coloured sea lily

Global distribution: Northeastern Atlantic; from the northern Bay of Biscay to Madeira and south of the Canary Islands

Habitat: This benthic species occurs at a depth range of 900 - 2070 m

Biology: This species has 10-21 smooth arms with isotomous branching. Its length of the stalk varies between 3.5 and 22.5 cm and the middle and distal stalk are pentagonal to circular in cross section. There are 5 robust cirri and the proximal ones are oriented upward.

Bioluminescence: In this species, the light appeared as a general feeble glow. However, it is not luminescing readily. But on prolonged immersion in freshwater, a pale blue-green light appeared all over the arms and less bright light appeared at the internode region of each segment of the cirri. However there was no light emission from the stalk of the animal (Herring, 1974)

Monachocrinus aotearoa

*Common name (s):*Not known

Global distribution: Tasman Sea, E of Auckland, New Zealand

Habitat: This benthic species has a depth range of 2050-2150m

Biology: Not reported

Bioluminescence: In this species, light emission from its arm tissue has been reported to its luminescent organ *viz.* sacculi. Further, the bioluminescence of this species is based on coelenterazine-luciferase system (Pablo,2020); Herring,1974).

Neocrinus decorus

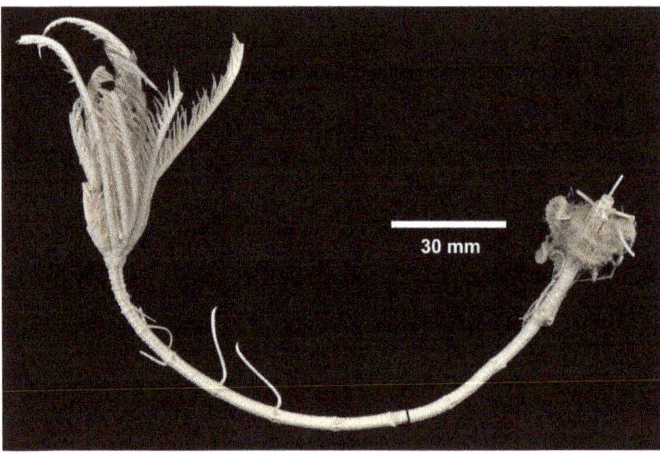

Image credit: Creator Eric A. Lazo-WasemGall L, Wikimedia Commons

Common name (s): Sea lily

Global distribution: Caribbean of Venezuela; West Indies

Habitat: This sessile species lives at depths from 154 to 1219 meters on hard substrate. Adult is found fixed to the sea bottom by a stalk.

Biology: The sea lily is resembling a plant and it has a stalk which is surmounted by a bulbous body with frondlike tentacles. The stem consists of limy disks, and the body has an internal skeleton of close-fitting limy plates. This species has a maximum length of 60 m.

Bioluminescence: This species has been reported to produce flashes of light along its stalk (Johnsen *et al.*, 2012)

Thalassometra lusitanica

Image credit: Charles G. Messing, Reproduced with permission

Common name (s): Not reported

Global distribution: This species is widely distributed at all latitudes; common in Off Morocco

Habitat: It is a deep-sea species

Biology: It is an yellow-colored species. All the species of this genus are characterized by having cirri which are long and slender with more than 25 segments. Distal cirrals are much shorter than proximal cirrals; and are broader than long with dorsal spines or processes. There are 5 radials and 10-15 arms which are aborally rounded.

Bioluminescence: In this species, light emission from its arm tissue has been reported to its luminescent organ *viz.* sacculi. Further, the bioluminescence of this species is based on coelenterazine-luciferase system (Pablo,2020); Herring,1974). Herring (1995) reported that it emitted a blue-green light with λ maximum at 485 nm. Further, light of this species was from a single region in each segment and it was not visible on the main arms.

Thaumatocrinus jungerseni

Image credit: Charles G. Messing, Reproduced with permission

Common name (s): Not known

Global distribution: Denmark Strait and SW of Iceland

Habitat: This deep-sea species occurs at a depth range of 823-2,075 m.

Biology: The species of this genus are characterized by the presence of 10 undivided arms arising from ten radials.

Bioluminescence: The light of this species appeared as a sparkling point. Further the light emission of this species has been reported to be due to luciferin/luciferase reaction and there was neuronal control of luminescence. Herring (1974) reported that freshly caught specimens of this species gave out a common pale greenish-blue glow which was brightest at the disc. This glow spread up each arm when the animal was handled. Further it was also reported that addition of hydrogen peroxide developed a great increase in the intensity of glow. This peroxide-stimulated animals showed that the light was produced at a series of point sources on each segment of the arms and cirri. Furthermore, it was also reported that the addition of acetylcholine and esterine produced prolonged glows with latencies of 10-20 sec. (Herring,1974)

3.8.4. Holothuroids

Class: Holothuroidea

Benthodytes sp.

Common name (s): Not reported

Global distribution: It is found in oceans all over the world.

Habitat: It is a deep-sea demersal nektonic species preferring sand substrata and is largely found at depths of about 1900m

Image credit: Smithsonian National Museum of Natural History. (Applied for permission)

Biology: Body of the species of *Benthodytes* is more or less depressed. Mouth is ventral, and is at a greater distance from the foremost extremity of the body. Anus is posterior, dorsal and is almost terminal. Tentacles are 12-20. Pedicels are arranged in a single row round the brim of the body. Dorsal surface is seldom naked, and is commonly with a number of retractile or non-retractile, processes.

Bioluminescence: In all the species of sea cucumbers, the light appear from granular cells of several discrete points which are believed to be more concentrated over the dorsal surface of the body and particularly at the distal tips of protuberances if any (Herring,1974). Emson *et al.* (1995a) reported that in the species of *Benthodytes,* the luminous sources are present at the bases of the papillae.

Benthogone rosea

Image credit: EOL (CC/NC). (Applied for permission)

Common name (s): Not reported

Global distribution: Eastern Atlantic

Habitat: This deep-sea, benthic species is found on the middle and lower continental slope with fine sands and highest contents of organic matter. It has a depth range of 1170m. 1400-1800m

Biology: Body of this species is bilaterally symmetrical.

Bioluminescence: In this bioluminescent species, minute blue-green point sources appeared from the whole dorsal body surface when the animal was treated in freshwater (Herring,1974)

Elpidia glacialis

Image credit: WoRMS. (CC by NC). (Applied for permission)

Common name (s): Cold Water Sea Cucumber; Sea pig

Global distribution: Southwest Pacific, Northeast Atlantic, Arctic and Antarctic.

Habitat: It. occurrs on the lower slopes of the continental shelf at depths of about 2,700 metres

Biology: It has enlarged tube feet that have taken on a leg-like appearance, These legs, in conjunction with their large, plump appearance have taken the common name "sea pig". It attains a maximum length of 15 cm.

Bioluminescence: It emits a bright blue-green light from the tips of the papillae. Luminescence is brightest on the tentacles and round the mouth (Emson *et al.*, 1995a)

Enypniastes eximia

Common name (s): Headless chicken fish, headless chicken monster, Spanish dancer. swimming sea cucumber, and pink see-through fantasia.

Enypniastes sp.

Image credit: Wikipedia

Global distribution: It is found all over the globe in many different regions. It is a rominent member of the benthic boundary layer community in deep Caribbean waters

Habitat: It is a benthopelagic species inhabiting at a depth of 1000 m

Biology: It has a round bulbous body, bifurcated tentacles, and a large anterior sail. Webbed swimming fin like structures are seen at the front and back of is body. Its most distinct feature is its coloration. While small enypniastes are bright pink, larger individuals are more reddish-brown color. It is also semi-transparent. It has a size range of 11-25 cm

Bioluminescence: In this species, its blue-green light production (λ maximum at 474nm) is triggered mechanically, and is produced by hundreds of granular bodies (20-60 um dia) present within the gelatinous integument of the animal. Its integument is quite fragile, and strong physical contact readily causes the skin of the animal o be sloughed off in a glowing cloud. The degree of luminous response of this species is therefore a function of the severity of contact. In the dark, near-bottom habitat, physical contact by a predator is believed to elicits light production in this species (Robinson,2009; Emson *et al.,* 1995a). Johnsen *et al.* (2012) reported that its light production at 470nm (Lambda max) consisted of adhesive luminescent secretions (or possibly detached luminescent epidermis).

Galatheathuria sp.

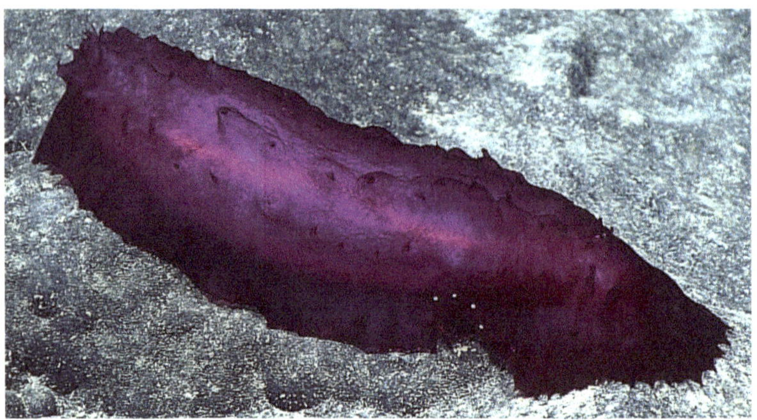

Image credit: NOAA Okeanos Explorer, Wikimedia Commons

Common name (s): Swimming sea cucumber

Global distribution: Central Pacific Ocean; South China Sea

Habitat: Thjs bathypelagic species occurs in abyssal plain at depths of 3400-3800 m.

Biology: It is a large free-swimming holothurian. Body is broad and oval, and is divided into a main body with a lateral brim (by undulations of which the animal is supposed to swim) and a head-like fore-end without a swimming brim. Mouth is central on the ventral side of the fore-end. Tentacles are rather dendritic and are completely retractile. This animal swims by the undulations of its lateral brim

Bioluminescence: In this pelagic unidentified species, the luminescence is more evenly distributed over the entire animal's surface. It has been reported to luminesce with a blue-green light from a multitude of minute points all over the dorsal and ventral body wall. This animal emits light both by handling and by immersing in freshwater (Herring,1974)

Hansenothuria benti

Common name (s): Not reported

Global distribution: Western Atlantic; Bahamas and other Caribbean islands.

Habitat: This upper bathyal, epibenthic species has a depth range of 363 - 904 m

Biology: It is a subcylindrical species with a moderate size of 13-23 cm in total length. It is approximately four to five times as long as broad. Body is fragile with light blue to pale purple coloration. Body wall is thick, gelatinous, and transparent. Anterior and posterior ends are gently tapering. While the anterior end is high and bluntly rounded, posterior end is low, and is narrowed to form short tail. Ventral surface is a flattened sole with minute, hairlike tube feet. Papillae rare regularly spaced in slightly zigzag rows, and are extending along the length of the body.

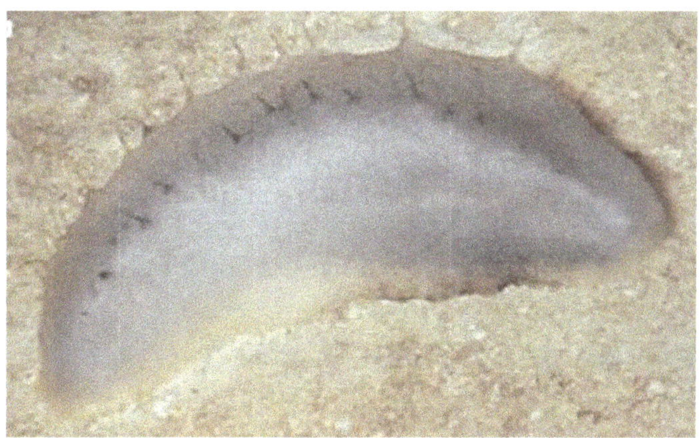

Image credit: John E. Miller and David L. Pawson. (Applied for permission)

Bioluminescence: In this species, its dorsal papillae are bioluminescent. Its emission peak wavelength has been reported to be 475 nm (lambda max) (Johnsen *et al.*, 2012)

Kolga hyalina

Image credit: Bodil Bluhm (UiT The Arctic University of Norway) and Katrin Iken (University of Alaska Fairbanks), with support from NOAA funding, Reproduced with permission

Common name (s): Arctic sea cucumber

Global distribution: Northeast Atlantic Ocean. It is commonly occurring in the Central Arctic Ocean.

Habitat: This benthic species has a depth range of 652 - 6235 m

Biology: It is a small animal of about 3 mm long with bilateral symmetry.

Bioluminescence: It has been reported to emit a bright blue light all over the tentacles when it is immersed in hydrogen peroxide. Under a binocular microscope, each minute point of light flared up and then died away quickly (Herring,1974)

Laetmogone violacea

Image credit: Ismiliana Wirawati, Reproduced with permission

Common name (s): Not known

Global distribution: Temperate Antarctic Atlantic: Weddell Sea, East Antarctic.

Habitat: This bathyal species has a depth range of 225 - 2500m

Biology: It is a mud eating sea cucumber with bilateral symmetry.Body of this species is elongated with length of 140 mm and width of 60 mm. Body coloration is brownish to pale-grey. There are 15 tentacles which are peltate and relatively large. Dorsal body wall is very thin and wrinkled. Dorsal papillae are very long, slender, and are arranged in a single row along each ambulacral groove. Tube feet are arge and rather bulky on each side of the brim. Mouth is ventral, and anus is terminal in position.

Bioluminescence: When threatened, it exhibits bioluminescence. Further, when it is immersed in freshwater, it emitted a brilliant blue-green light from the tips of all the dorsal papillae. Several smaller brightly luminous points also appeared from the rest of the dorsal and ventral body surface. Furthermore, its lateral body wall also emitted general flashes (Herring,1974)

Paelopatides sp.

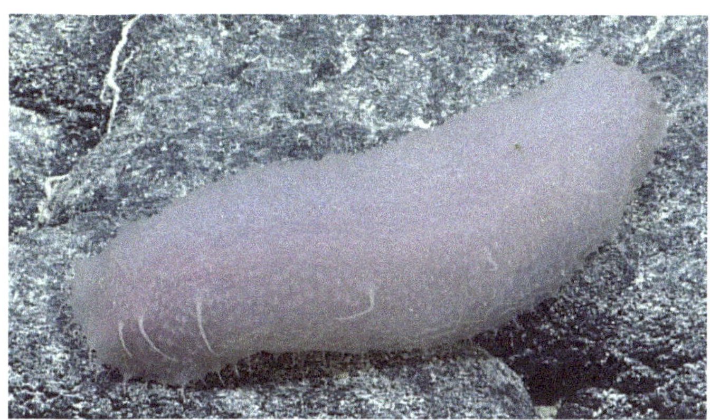

Image credit: NOAA Ocean Exploration. (PD), Reproduced with permission

Common name (s): Hadal Sea Cucumber

Global distribution: Mariana Trench;

Habitat: It occurs at a depth of about 6500m. It is known to live in a frozen temperature of 1.7° C in hadal trenches

Biology: Body coloration of the species of this genus is translucent white or light brown. Individuals are up to 330/ mm in length, and are with 19 pairs of dorsal appendages. Ventral side has one ventral ambulacra, with two rows of tube feet. Ossicles are present in dorsal appendages, tube feet, tentacles and body wall near the anus.

Bioluminescence: In the specimen of this unidentified species, athin sprinkling of blue-green point sources was oserved when it was treated with freshwater. Further a pattern of brighter spots appeared in four lines (two laterally and two dorsally) which were running along the entire length of the animal (Herring,1974)

Pannychia moseleyi

Common name (s): Not reported

Global distribution: It has cosmopolitan distribution

Habitat: It occurs in the benthic zone areas such as soft sediment, rock outcrops, lava at a depth range of 212 - 2599 m

Image credit: NOAA/MBARI, Wikimedia Commons

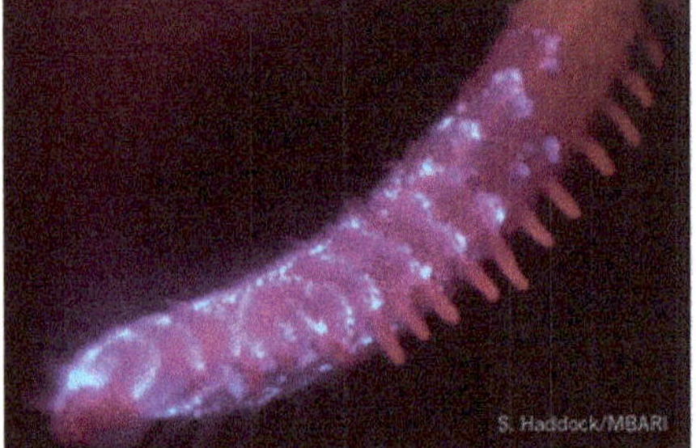

Left: Animal; Right: Bioluminescence

Image credit: Monterey Bay Aquarium Research Institute. (Applied for permission)

Biology: Animals of this species are subcylindrical dorsally and laterally; and flat ventrally. While mouth is anterior and ventral, anus is posterior and terminal. There are 20 tentacles. A single series of 29–30 tube feet is present on each lateroventral ambulacrum, and irregular double series of 55 smaller tube feet mid-ventrally. Color of the preserved animal is white grey; and dark violet dorsally. Ends of tentacles and tube feet are yellowish. Size of the animals may be up to 200 mm long and 40 mm wide.

Bioluminescence: This species has been reported to produce bioluminescence in the form of waves of blue and green light travelling along its entire body. (https://en.wikipedia.org/wiki/Pannychia_moseleyi). Emson *et al.* (1995a) and Brian (1992) reported that this species elicited single or multiple blue-green spiral and quasi-circular waves of luminescence both anteriorly and posteriorly *i.e.* over the entire body surface except the podia, papillae, and tentacles

Paroriza sp.

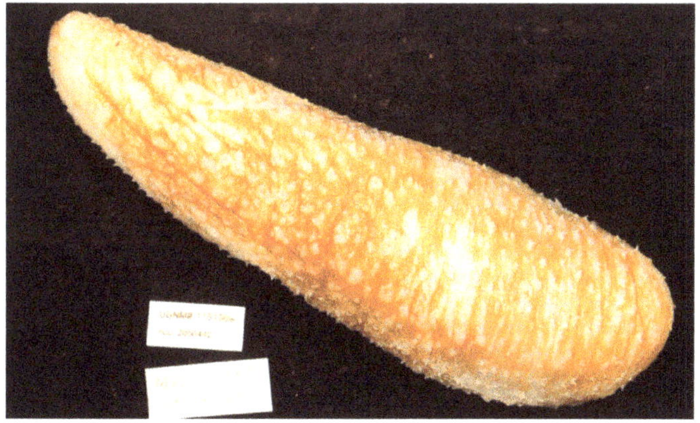

Image credit: Allen Collins, Wikimedia Commons

Common name (s): Not reported

Global distribution: North-east Atlantic

Habitat: It is a deep-sea deposit-feeding holothurian

Biology: Body of this unidentified species has a cream coloured leathery integument.

Bioluminescence: Herring (1974) reported that the leathery integument of this species emittted a brilliant blue-green light from the two paired lines of the small dorsal papillae. Further, feebler light was also found emitted from the points of the general body wall.

Pelagothuria natatrix

Image credit: NOAA Okeanos Explorer Program, Galapagos Rift Expedition 2011, Wikimedia Commons

Common name (s): Bottom bouncer

Global distribution: Atlantic, Pacific and Indian oceans; it is more common in Eastern Pacific. However, it has also been reported to live all over the world.

Habitat: This deep sea, swimming sea cucumber has a common depth range of 0 - 540 m. However, it has been reported to live in the depth range of 200-6000 m. This one has been reported to spend its entire life swimming through the sea and need never touch the bottom.

Biology: It's the only truly pelagic and passive swimming echinoderm in the world. It looks more like a jellyfish with its large umbrella-like swimming structure supported by a ring of around 12 highly modified oral tentacles. Apparently these oral tentacles can be extended and spread out. It has a small tapered body and its swimming position is with the mouth on top. Body is translucent with a pale purple pigmentation. Mouth is surrounded by 15 short feeding tentacles. It reaches around 16 cm in total diameter.

Bioluminescence: Emson *et al.* (1995a) reported that its whole body is weakly luminous at 472nm, λ maximum.

Peniagone hollisi

Image credit: Adrian Glover, Reproduced with permission

Common name (s): Not known

Global distribution: Northwest Pacific Ocean. Commonly occurs in all oceans except for the Arctic, with the highest species diversity in the Antarctic and the Pacific Ocean

Habitat: It is found distributed from 220 to 8660 m.

Biology: Not reported

Bioluminescence: It showed a bright blue-green light (λ maximum at 463 nm) from the tips of the papillae. Further, luminescence was found to be brightest on the tentacles and round the mouth (Emson *et al.*, 1995a)

Penilpidia ludwigi (= *Irpa ludwigi*)

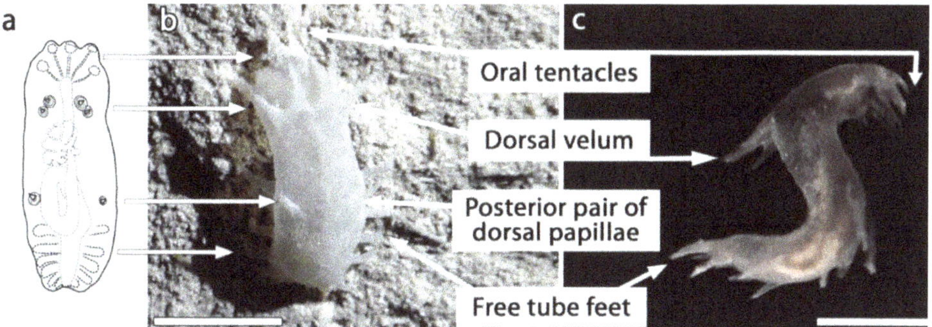

Distribution and Swimming Ability of *Penilpidia ludwigi*

Image credit: G. Chimienti1,2 and R. Aguilar3 and A. V. Gebruk4 and F. Mastrototaro (Applied for permission)

Common name (s): Not known

Global distribution: Western, Central, and Eastern Mediterranean Sea

Habitat: It lives above the seabed in sediment traps.It is capable of swimming in deep-sea habitats including seamounts, canyons, and ridges at depths from 1200 to 1700m.

Biology: The specimens of this species are generally unpigmented and transparent with the exception of the greyish-brown digestive tract, which was due to the sediment contained in it. Body shape is elongated, ovoid and dorso-ventrally flattened. Mouth is surrounded by 10 large tentacles forming a mouth tube. Posterior half of the body bears six pairs of free tube feet.

Bioluminescence: Its body has been reported to be weekly luminous and intense light (λ maximum at 463 nm) was found emitted from all the 6 dorsal papillae and the postero-lateral podia (Emson *et al.*, 1995a)

Psychropotes depressa (= *Euphronides depressa*)

*Common name (s):*Not known

Global distribution: It has cosmopolitan distribution; North and South Atlantic, and Pacific: off Japan, Caroline seamount, Gulf of Panama and Chile.

Habitat: This deep-sea swimming species lives at a depth range of 957–4200 m.

Biology: Body of this species is convex dorsally and flattened ventrally. Colour of body is pinkish both dorsally and ventrally; and internal organs are transparent and are clearly visible through body wall. Skin is rather thin. There are 18 tentacles which are in a single circle, non-retractile, and with round terminal discs. Midventral tubefeet are small and there are 4 pairs of dorsal papillae in 2 rows. The maximum length and width of the animal are 200 mm and 47 mm respectively.

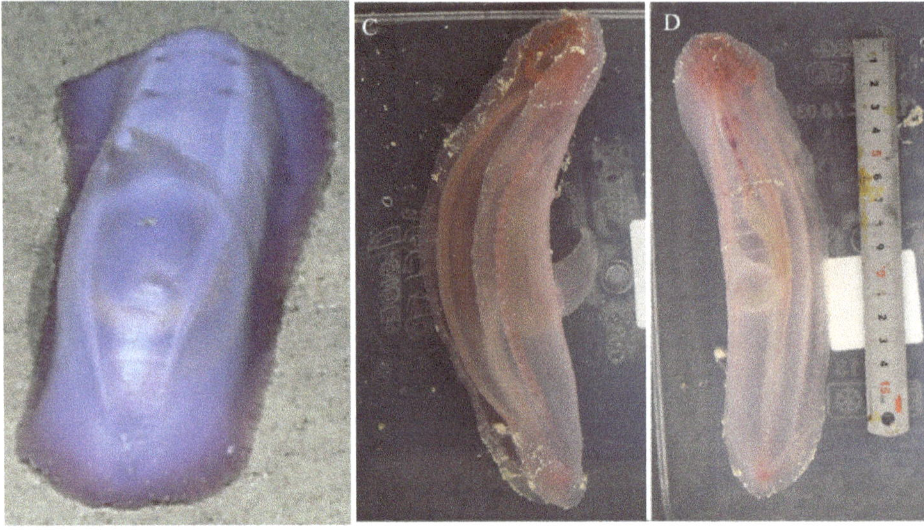

Image credit: Smithsonian National Museum of Natural History (Applied for permission)

Image credit: Ning Xiaoa, Xiaomeng Lic and Zhongli Shaa (Applied for permission)

Bioluminescence: In this species, the intense light regions are seen on sail and/ or dorsal papillae. When the animal was immersed in freshwater, it sparkled with points of blue light all over the dorsal surface of the body. (Herring,1974)

Psychropotes longicauda

Image credit: NOAA Okeanos Explorer, Wikimedia Commons

Common name (s): Gummy squirrel

Global distribution: It has cosmopolitan distribution; Indo-West Pacific and Antarctic Indian Ocean: Kerguelen and South China Sea.

Habitat: This abyssal species is found on soft sediments at a depth range of 1100 - 5173

Biology: Body of this species is very flexible and approximately cylindrical. But it is broadest and somewhat flattened near the anterior end. At the posterior end, there is a dorsal, tail-like appendage which is up to 12 cm long, and is either pointed or with a pair of unequal-length tips. There is a single row of larger tube feet on each side of the body and a double row of small tube feet along the ventral surface. Skin is soft and pliable and the colour is violet in alcohol- preserved specimens. This species can grow to a length of 14 - 32 cm.

Bioluminescence: Pablo (2020) reported that the caudal skin tissue of this species is epifluorescent due to the presence of granular cells. Further he suggested on the presence of coelenterazine-luciferase system in this species

Scotoplanes gibbosa

Image credit: NOAA/MBARI, Wikimedia Commons

Common name (s): Sea pig

Global distribution: It is found distributed in almost all deep-sea regions in the world. It is commonly found off the coast of San Diego, Arctic, Atlantic, Pacific, and Indian Oceans.

Habitat: This deep-sea species is often observed on the ocean floor in groups of 10 to 30 and 600 individuals have also been observed in one congregation. Its depth range is 1000-10000m

Biology: Body of this species is translucent with white coloration. It is bilaterally symmetrical, and is covered in tube-like feet which are found on top of the animal are used in locomotion and possibly respiration. It is typically 2 to 15 cm in length.

Bioluminescence: It showed a bright blue-green light (λ maximum at 463 nm) from the tips of the papillae. Luminescence has also been reported to be the brightest on the tentacles and round the mouth. In freshwater treatment, the animal was however, weakly luminous (Emson *et al.*, 1995 a)

Chapter 4

Industrial and Biomedical Applications of Bioluminescent Marine Invertebrates

Bioluminescence has been reported to possess several applications for use in industry, daily life or in a clinic. In the not so distant future, our traditional street lamps may be replaced by glowing trees and buildings. The industrial and biomedical applications such as ATP sensing, hygiene control in the fish and milk industries, mapping pollution in ecosystems using bioluminescence based assays, high-throughput screening in drug discovery and in vivo imaging of tumours as well as infections are widely discussed nowadays. In this chapter the potential applications of marine bioluminescence in general and that of marine luminescent invertebrates are discussed.

1. ATP Bioluminescence Assay and Hygiene Control

Surgical site infections (SSIs) have been reported to represent about 20 per cent of all Healthcare Associated Infections (HAIs) in many developed countries. These SSIs lead to higher morbidity and mortality of surgical patients and increase the healthcare cost considerably. Further, such contaminated environments play a significant role in the transmission of pathogenic microorganisms, including multidrug-resistant organisms. Operating rooms are therefore considered to be very high-risk areas, where high hygiene standards need to be constantly guaranteed.

Though traditional microbiological techniques are considered to be the most commonly used methods to evaluate hygienic quality, they require specific skills, long execution and analysis times and are therefore unsuitable for routine monitoring. In the last decade, alternative methods like adenosine triphosphate (ATP) bioluminescence assay proposed for assessing environmental cleanliness. This ATP bioluminescence assay is based on the measurement of levels of ATP present on an environmental surface. This bioluminescence test utilizes the chemiluminescence properties of luciferin-luciferase reagent, which reacts with any ATP residue present on a substrate, emitting light and measuring the presence and level of organic matter as contaminants. This easy to use ATP-bioluminescence which was first applied in the food industry provided rapid results and it has already been investigated as an objective tool to assess hospital cleanliness even in high-risk areas. Further, ATP is considered to be an indicator of the presence of organic material rather than microbial contamination. Although some studies have found a significant correlation between ATP levels and TVC (Total Viable Counts) values, a direct relationship remains controversial and few studies have expressed their doubts about the utility of bioluminescence for hygiene monitoring especially within healthcare settings. Keeping these in consideration, Sanna *et al.* (2018) assess the feasibility of using the bioluminescence technique as a rapid analytical method directly in situ to verify the hygienic quality of the most critical surfaces in operating theatres and the effectiveness of sanitization procedures. According to these authors, the examined surfaces presented very low TVCs. Further, the ATP mean values showed a progressive increase from surfaces with TVC = 0 to surfaces with TVC > 15 colony forming units (CFU)/plate. These results suggested that the ATP-bioluminescence-assay can be a useful tool to measure the efficiency of cleaning procedures in environments with very low microbial counts

2. Bacterial Bioluminescent Technology to Food Microbiology

Bioluminescent Lux-tagged bacteria such as *Escherichia coli and Listeria* may be used for problem-solving and for the development of modified and improved processing (*e.g.* post-processing in yoghurt and cheese) and storage purposes (Ohmiya,http://photobiology.info/Ohmiya.html).

3. Mapping Pollution in Ecosystems

The whole-cell bacterial bioluminescence sensors are the most widely used bioluminescence sensors in the toxicology monitoring of ecosystems. Like most bioluminescence-based assays, these sensors offer a quick result to help assess toxicity levels. The benefit of bioluminescent bacteria is that their light output can be employed as a rapid measure of survival. These bacterial bioluminescence sensors have been used to monitor a wide variety of contaminants including both heavy metal contaminants and organic compounds (alcohol, carboxylic acids and aromatic compounds). For example bioluminescence from *Aliivibrio fischeri* is used nowadays to monitor water toxicity. When exposed to pollutants, light output from this bacterial culture decreases, signalling the possible presence of a contaminant (https://theconversation.com/what-is-bioluminescence-and-how-is-it-used-by-humans-and-in-nature-100472).Further, bioluminescent bacteria are also being used to create unique glowing artwork and proposals are on the way to use them in indoor aquariums as part of aesthetic architecture in skyscrapers (Syed and Anderson, 2021).Some bacterial bioluminescence biosensors used in ecotoxicology and their detection limits are given below:

Use of Bacterial Bioluminescence Biosensors in Ecotoxicology

Target	*Bacterial Species*	*Detection Limit*
Arsenite	*Escherichia coli*	39.6 mg L^{-1}
Terbutryn (Herbicide)*	*Aliivibrio fischeri*	81 mg L^{-1}
Mercury	*Photobacterium leiognathi*	9.87 mg L^{-1}
Chlorine	*Escherichia coli* mutants	1 mg L^{-1}
Sucralose (sweetener)	*Escherichia coli* mutants	1 g L^{-1}

Source: Syed and Anderson, 2021.

4. Role of Bioluminescence in Cell Biology and Disease Dynamics

The medical applications of bioluminescence have attracted the most excitement. The discovery and development of green fluorescent protein (GFP) from the crystal jellyfish Aequorea Victoria assumes greater importance. Unlike the bioluminescence mechanism described so far for other luminescent organisms, it is fluorescent. It means the protein of GFP needs to be excited by blue light before emitting its characteristic green light. This GFP has been genetically inserted into various cell types and even animals and this has shed light on important aspects of cell biology and disease dynamics. Further, this GFP and other fluorescent proteins have revolutionized research in fields from immunology to neuroscience. Furthermore, the illumination power given by these fluorescent proteins is transforming most areas of modern biomedicine (Dybas,2019; https://theconversation.com/what-is-bioluminescence-and-how-is-it-used-by-humans-and-in-nature-100472)

5. Medical Applications of Bioluminescence: Marine Bioluminescence in Cancer Immunotherapies

Bioluminescence has been reported to play a significant role in human health through its participation in "Immunoassays, gene expression assays, drug screening, bioimaging of live organisms, cancer studies and the investigation of infectious diseases"(Dybas,2019). The glimmering power of bioluminescence has been harnessed for lifesaving uses in medicine, from lighting up structures inside the brain to illuminating the progression of cancer cells. Bioluminescent systems are used in the imaging of developing tumours in cell studies. Although initial studies related to the monitoring of tumours and gene expression subcutaneously, more recent work has focused on the imaging of more-deep seated tumours and challenging targets such as imaging in the brain of moving animals with novel red-shifted and bright luciferin–luciferase pairs.

Recent studies have portrayed the role of marine bioluminescence in treating several cancers, including chronic myelogenous leukemia, acute myelogenous leukemia, and Burkitt lymphoma. For example when, the Matador assay-Luciferase from the bioluminescent Pacific Ocean crustaceans was introduced into cancer cells, the enzyme leaked out as the cancer cells died, leaving a visible glow that can be measured. This assay could accurately recognize the death of a single cancer cell. Further, as the Matador assay was able to detect cell death in 30 minutes, it will lead to faster treatments for patients getting presently immunotherapies such as CAR-T cells. (Chimeric antigen receptor (CAR) T-cell therapy is a way to get immune cells called T cells (a type of white blood cell) to fight cancer by changing them in the lab so they can find and destroy cancer cells). Furthermore, Matador assay can be employed in as little as half an hour. Therefore, this has major uses not only in research on next-generation CAR-T cell therapies but also on CAR-T cells that are in current clinical use (Dybas,2019)

Bioluminescence Imaging and Infectious Diseases

Bioluminescent Imaging is also presently used to study the interaction of infectious microorganisms with living cells. This imaging has also been effectively used in imaging the development of infectious diseases both in vitro and in vivo. Genetically modified bioluminescent pathogens, such as bacteria, parasites, viruses and fungi have been designed and monitored both in vivo and in vitro, in the presence and absence of therapies to test their effectiveness. The bioluminescent light output has been shown to correlate with the infection load.

References

Anderson, F.E., Bergman, A., Cheng, S.H., Pankey, M.S., and Valinassab, T. (2013). Lights out: the evolution of bacterial bioluminescence in Loliginidae. *Hydrobiologia* 725: 189-203.

Berry, S.S. (1920). Light Production in Cephalopods, I. An Introductory Survey, *Biol.l Bull.* 38: 141-169.

Bessho Uehara, M., Francis, W.R., and Haddock, S.H.D. (2020). Biochemical characterization of diverse deep sea anthozoan bioluminescence systems., *J. Mar. Biol.* (2020) 167:114.

Bilbaut, A. (1980). Excitable epithelial cells in the bioluminescent scales of a polynoid worm; Effects of various ions on the action potentials and on the excitation-luminescence coupling. *J. exp. Biol.* 88: 219-238.

Birk, M.H., blichef, M.E., and Garm, A. (2018). Deep-sea starfish from the Arctic have well-developed eyes in the dark. *Proc. R. Soc. B* 285: 20172743.

Brehm, P., and Morin, J.G. (1977). Localization and characterization of luminescent cells in *Ophiopsila californica* and *Amphipholis squamata* (Echinodermata: Ophiuroidea). *Biol Bull.* 152:12-25.

Brian, A. (1992). Bioluminescence in the benthic holothurian *Pannychia moseleyi.*, https://searchworks.stanford.edu/view/10773078, 23p.

Brugler, M.R., Aguado, M.T., Tessler, M., Siddall, M.E. (2018). The transcriptome of the Bermuda fireworm *Odontosyllis enopla* (Annelida: Syllidae): A unique luciferase gene family and putative epitoky-related genes. PLoS ONE 13 : e0200944.

Chan, B.K.K., Lin, I., Shih, T., and Chan, T. (2008). Bioluminescent Emissions of the Deep-Water Pandalid Shrimp, *Heterocarpus sibogae* De Man, 1917 (Decapoda, Caridea, Pandalidae) under Laboratory Conditions. *Crustaceana* 81: 341-350.

Chimienti, G., Angeletti, L., and Mastrototota, F. (2018). Withdrawal behaviour of the red sea pen *Pennatula rubra* (Cnidaria: Pennatulacea), *Eur.Zool.J.* 85: 64-70.

Clarke, G.L., Conover, R.J., David, C.N., and Nicol, J.A.C. (1962). Comparative studies of luminescence in copepods and other pelagic marine animals. *J.mar.biol.UK.* 42: 541-564.

Cohen, A.C., and Morin, J.G. (1989) Six new luminescent ostracodes of the genus *Vargula* (Myodocopida: Cypridinidae) from the San Blas region of Panama. *J.Crust.Biol.* 9: 297–340.

Cohen, A.C., and Morin, J.G. (1986).Three new luminescent ostracodes of the genus *Vargula* (Myodocopida, Cypridinidae) from the San Blas region of Panama. Contributions in Science, *Natural History Museum of Los Angeles County* 373: 1–23.

Cohen, A.C., and Morrin, J.G. (2017). Sexual Morphology, Reproduction and the Evolution of Bioluminescence in Ostracoda. *Virtual Paleontology* 9: 37-70.

Cohen, A.C. (2010). Two New Bioluminescent Ostracode Genera, *Enewton* and *Photeros* (Myodocopida: Cypridinidae), with Three New Species from Jamaica. *J.Crust.Biol.* 30:1-55.

Cohen, A.C., and Morin, J.G. (1993). The cypridinid copulatory limb and a new genus *Kornickeria* (Ostracoda: Myodocopida) with four new species of bioluminescent ostracods from the Caribbean. *Zool. J. Linn. Soc.* 108: 23-84.

Copila'-Ciocianu, D., and Pop, F.M. (2020). An account of bacterial–induced luminescence in the Ponto-Caspian amphipod *Pontogammarus maeoticus* (Sowinskyi, 1894), with an overview of amphipod bioluminescence. *North-West. J. Zool.* 16: 238-240.

Cormier, M.J. (1962). Studies on the Bioluminescence *of Renilla reniformis*. J.*Bio. Chem.* 237: 2032-2037.

Dales, R.P. (2011). Bioluminescence in Pelagic Polychaetes. 2011, *J. Fish. Res.* 28:1487-1489.

Davenport, D., and Nicol, J.A.C. (1956). Observations on luminescence in sea pens (Pennatulacea). *Proc. R. Soc. Lond. B* 144: 480–496.

Deheyn, D., Mallefet, J., and Jangoux, M. (2000). Expression of bioluminescence in *Amphipholis squamata* (Ophiuroidea: Echinodermata) in presence of various organisms: a laboratory study. *J.mar.biol.Ass., UK.*, 80: 179-180.

Deheyn, D., Mallefet, J., and Jangoux, M. (1999).Variation in bioluminescence with ambient illumination and diel cycle in a cosmopliran ophiuroid (Echinodermata). *Cah.Biol.Mar.* 40: 57-63.

Deheyn, D.D., and Latz, M.I. (2009). Internal and Secreted Bioluminescence of the Marine Polychaete *Odontosyllis phosphorea* (Syllidae). *Invertebr. Biol.* 128: 31-45.

Delroisse, J., Ullrich- luter, E., Blaue, S., Eeckhaut, I., Flammang, P., Mallefet, J. (2017). Fine structure of the luminous spines and luciferase detection in the brittle star *Amphiura filiformis*. *Zool. Anz.* 269: 1-12.

Delroisse, J., Ullrich-Lu¨ter, E., Blaue, S., Martinez, O.O., Eeckhaut, I., Flammang, P., and Mallefet, J. (2017a). A puzzling homology: a brittle star using a putative cnidarian-type luciferase for bioluminescence. *Open Biol.* 7: 160300.

Dewael, Y., and Mallefet, J. (2002). Luminescence in ophiuroids (Echinodermata) does not share a common nervous control in all species. *J Exp Biol.* 205: 799-806.

Ding, B., and Liu, Y. (2017). Bioluminescence of Firefly Squid via Mechanism of Single Electron-Transfer Oxygenation and Charge-Transfer-Induced Luminescence. *J. Am. Chem.* Soc. 139: 1106–1119.

Dorrestein, W.C.A., and Westheide, W. (2013). *Reproductive strategies and development patterns in Annelids.* Springer Science and Business Media, Science, 314p.

Dybas, C.L. (2019). Illuminating New Biomedical Discoveries: Bioluminescent, biofluorescent species glow with promise. *J. Biosci.* 69: 487–495,

Emmerson, W.D. (2017). *A Guide to, and Checklist for, the Decapoda of Namibia, South Africa and Mozambique.*Vol. 1, Cambridge Scholars Publishing, - Social Science, 590 p.

Emson, R., Smith, A., and Campbell, A. (1995a). *Holothuroid- Echinoderm Research.* CRC Press, Science, 341 p.

Emson, R.H., Smith, A.B. and Campbell, A.C. (eds) (1995) *Echinoderm Research* 1995. Proceedings of the Fourth European Echinoderms Colloquium, London, United Kingdom, 10-13 April 1995. A.A. Balkema, Rotterdam/ Brookfield, 341 p.

Emson, R.H., and Herring, P.J. (1985). *Bioluminescence in deep and shallow water brittlestars.* Echinodermata. CRC Press, eBook, ISBN9781003079224.

Foster, J.S., von Boletzky, S., and McFall-Ngai, M.J. (2002). A comparison of the light organ development of *Sepiola robusta* and *Euprymna scolopes* (Ccephalopoda: Sepiolidae). *Bull. Mar.Sci.*70 : 141-153.

Franci, W.R., and de Vilar, A.S. (2020). Bioluminescence and fluorescence of three sea pens in the north-west Mediterranean sea. *bioRxiv* preprint doi: https://doi.org/10.1101/2020.12.08.416396; 2020

Francis, W.R., Powers, M.L., and Haddock, S.H.D. (2016). Bioluminescence spectra from three deep-sea polychaete worms. *Mar.Biol.* 163, 255.

Frank, T.M., and Case, J.F. (1988). Visual Spectral Sensitivities of Bioluminescent Deep-sea Crustaceans. *Bio.Bull.* 175: 261.

Gaston, G., and Hall, J. (2000). Lunar Periodicity and Bioluminescence of Swarming *Odontosyllis luminosa* (Polychaeta: Syllidae) in Belize. *Gulf Caribb. Res.* 12 : 47-51.

Gerrish, G.A., and Morin, J.G. (2008). Life Cycle of a Bioluminescent Marine Ostracode, *Vargula annecohenae* (Myodocopida: Cypridinidae).*J.Crust.Biol.* 28 : 669–674.

Gouveneaux, A., Gielen, M., and Mallefet, J. (2017). Behavioural responses of the yellow emitting annelid *Tomopteris helgolandica* to photic stimuli. *Luminescence* 33, DOI:10.1002/bio.3440

Goodheart, J.A., Minsky, G., Brynjegard-Bialik, M.N., Drummond, M.S., Munoz, J.D., Fallon, T.R., Schultz, D.T., Weng, J., Torres, E., and Oakley, T.H. (2020). Laboratory culture of the California Sea Firefly *Vargula tsujii* (Ostracoda: Cypridinidae): Developing a model system for the evolution of marine bioluminescence. *Sci.Rep.* 10, 10443, https://doi.org/10.1038/s41598-020-67209-w

Goodwin, A. (2006). Use of *Renilla* Bioluminescence to Illustrate Nervous Function. *ABLE Proc.* 28::217-226.

Gotto, R.V. (1963). Luminescent Ophiuroids and Associated Copepods. *Ir. Nat.' J.* 14: 137-139.

Gouveneaux, A. (2016). *Bioluminescence of Tomopteridae species (Annelida): multidisciplinary approach.*Ph.D.Thesis, Université catholique de Louvain.

Grober, M.S. (1988). Brittle-star bioluminescence functions as an aposematic signal to deter crustacean predators. *Anim.Behav.* 36: 493-501.

Haddock, S.H.D., Moline, M.A., and Case, J.F. (2010). A Bioluminescence in the Sea. *Ann. Rev.Mar. Sci.* 2: 443-493.

Haddock, http://photobiology.info/Haddock.html

Haddock, http://photobiology.info/Haddock.html; https://biolum.eemb.ucsb.edu/organism/

Haneda, Y. (1963). Observation on the luminescence of the shallow water squid, *Uroteuthis bartschi. Sci.Rep. Yokosuka City Museum.*8: 10-16.

Harvey, E.N. (1917). Studies on Bioluminescence.VI. Light Production by a Japanese Pennatulid, *Cavernularia haberi.* https://doi.org/10.1152/ajplegacy.1917.42.2.349

Heffing, P.J. (1974). New observations on the bioluminescence of echinoderms.*J.Zool.* 172: 401-418.

Heger, A., King, N.J., Wigham, B.D., Jamieson, A.J., Bagley.P.M., Allan, L., Pfannkuche, O., and Priede, I.G. (2007). Benthic bioluminescence in the bathyal North East Atlantic: luminescent responses of *Vargula norvegica* (Ostracoda: Myodocopida) to predation by the deep-water eel (*Synaphobranchus kaupii*), *Mar. Biol.*151:1471–1478.

Henry, J., and Ninio, M. (1978). Control of the Ca2+-triggered bioluminescence of *Veretillum cynomorium lumisomes. Biochim Biophys Acta (BBA) - Bioenergetics* 504: 40-59.

Herring, P.J. (1995). Bioluminescent echinoderms: unity of function in diversity of expression. In *Echinoderm Research* (eds RH Emson, AB Smith, AC Campbell), pp. 9–17. Rotterdam, The Netherlands: A. A. Balkema.

Herring, P.J., and Barnes, A.T. (1976). Light-Stimulated Bioluminescence of *Thalassocaris crinita* (Dana) (Decapoda, Caridea). *Crustaceana,* 31: 107-110.

Herring, P.J. (1987). Systematic distribution of bioluminescence in living organisms. *J. biolumin. chemilumin.* 1: 147–163.

Herring, P.J. (1974). New observations on the bioluminescence of echinoderms. *J.Zool.* 172: 147-163.

Herring, P.J. (1981). The comparative morphology of hepatic photophores in decapod Crustacea. *J.Mar.Biol.Ass.UK.* 61: 723-737.

Herring, P.J., Widder, E.A., and Haddock, S.H.D. (1992). Correlation of bioluminescence emissions with ventral photophores in the mesopelagic squid *Abralia veranyi* (Cephalopoda: Enoploteuthidae). *Mar.Biol.* 112: 293-298.

Hill, M.R., and Begley, M. (2013). Shining light on food microbiology; applications of Lux-tagged microorganisms in the food industry. Trends *Food Sci. Technol.*32:14-15.

https://en.wikipedia.org/wiki/Savalia_lucifica

http://ecoursesonline.iasri.res.in/mod/page/view.php?id=86809)

http://bioweb.uwlax.edu/bio203/2011/kramolis_kali/

http://dsg.mbari.org/dsg/view/concept/Japetella%20diaphana

http://dsg.mbari.org/dsg/view/concept/Octopoteuthis%20deletron

http://ecoursesonline.iasri.res.in/mod/page/view.php?id=42450

http://ecoursesonline.iasri.res.in/mod/page/view.php?id=86809

http://photobiology.info/Haddock.html

http://species-identification.org/species.php?species_group=zsao and id=2647

http://species-identification.org/species.php?species_group=zsao and menuentry=inleiding and record=Po.%201%20

http://www.seaslugforum.net/showall/plocceyl

http://www.thecephalopodpage.org/Tdanae.php

https://animaldiversity.org/accounts/Stoloteuthis_leucoptera

https://animaldiversity.org/accounts/Vampyroteuthis_infernalis/

https://australian.museum/blog/amri-news/fire-shooting-butterfly-bobtail-named-in-honour-of-professor-merlin-crossley/

https://australian.museum/learn/animals/molluscs/southern-bobtail-squid-euprymna-tasmanica-pfeffer-1884

https://biolum.eemb.ucsb.edu/organism/

https://blog.biomall.in/bioluminescence-emission-of-light-from-a-living-organism/

https://cals.cornell.edu/news/cornell-discovered-shrimp-makes-2019-top-10-not-taste

https://chapinthehat.com/sff-video

https://collections.museumsvictoria.com.au/species/8642

https://conxemar.com/en/neon-flying-squid

https://cypridinidae.myspecies.info/taxonomy/term/74/descriptions

https://dailynexus.com/2020-07-23/bright-ideas-ucsb-researchers-establish-a-laboratory-culture-of-bioluminescent-ostracods/

https://doi.org/10.1515/9781400875689-032

https://ecology.wa.gov/Blog/Posts/March-2016/Eyes-Under-Puget-Sound-Critter-of-the-Month-%E2%80%94-Slen

https://en.wikipedia.org/wiki/Colossendeis

https://en.wikipedia.org/wiki/Colossendeis; Haddock, *et al.*, 2010

https://en.wikipedia.org/wiki/Euprymna_scolopes

https://en.wikipedia.org/wiki/Hinea_brasiliana

https://en.wikipedia.org/wiki/Iridoteuthis_merlini

https://en.wikipedia.org/wiki/Japetella

https://en.wikipedia.org/wiki/Midwater_squid

https://en.wikipedia.org/wiki/Neon_flying_squid). Anon. (https://conxemar.com/en/neon-flying-squid

https://en.wikipedia.org/wiki/Ptilosarcus_gurneyi

https://en.wikipedia.org/wiki/Sthenoteuthis_oualaniensis

https://en.wikipedia.org/wiki/Sthenoteuthis_pteropus

https://en.wikipedia.org/wiki/Swima_bombiviridis

https://eol.org/pages/14813/articles

https://irdb.nii.ac.jp/en/00822/0001821732

https://news.globallandscapesforum.org/57739/bioluminescence-it-requires-a-whole-new-way-of-seeing/

https://oceanexplorer.noaa.gov/explorations/05deepcorals/logs/nov12/media/movies/bioluminescing_bamboo_video.html

https://patents.justia.com/patent/20150064731

https://scholarspace.manoa.hawaii.edu/items/ab72894e-f208-4bc6-93c2-43c72d0791fc

https://sciencenordic.com/animals–plants-denmark-greenland-science-special/stunning-starfish-illuminates-the-dark-arctic/1454536

https://sumyuworld.wordpress.com/2012/02/06/deep-sea-beasties/

https://theconversation.com/what-is-bioluminescence-and-how-is-it-used-by-humans-and-in-nature-100472

https://today.ucsd.edu/story/bioluminescent_worm_found_to_have_iron_superpowers

https://www.australiangeographic.com.au/topics/science-environment/2018/08/bioluminescent-beauties-meet-australias-sparkling-species/

https://www.biographic.com/sucker-deception/

https://www.carolina.com/cellular-physiology-enzymes/sea-firefly-cypridina-hilgendorfii/FAM_203430.pr).

https://www.deviantart.com/monsieur-le-gris/art/Polycirrus-aurantiacus-179329719

https://www.google.com/search?q=bioluminescence+in+Chaetopterus and oq=bioluminescence+in+Chaetopterus and aqs=chrome.69i57j33i160l3.5469j1j15 and sourceid=chrome and ie=UTF-8)

https://www.marinelifephotography.com/marine/cnidaria/savalia-lucifica.htm

https://www.mbari.org/products/creature-feature/bomber-worm/

https://www.nhm.ac.uk/research-curation/scientific-resources/biodiversity/global-biodiversity/atlantic-ostracods/ostracods/structure/glands.html)

https://www.rawpixel.com/image/547488/bioluminescent-squid-vintage-poster

https://www.science.org/content/article/vomiting-shrimp-and-other-deep-sea-creatures-light-ocean-floor

https://www.sciencedaily.com/releases/2010/12/101214201534.htm

https://www.surg.org.au/species/nudibranchs-and-sea-slugs/polyceridae/plocamopherus/imperialis

https://www.wired.com/2012/01/glow-little-spewing-shrimp-glow/

https://australian.museum/learn/animals/molluscs/luminous-bay-squid-loliolus-noctiluca-lu-roper-tait-1985/

https://www.sealifebase.ca/summary/Taningia-danae.htmlh

Hoppe, B.W., Lohmann, W., Markl, H., Ziegler. H. (2012). *Biophysics.* Springer Science and Business Media - Science, 941 p.

Huber, M.E., Arneson, A.C., and Widder, E.A. (1989). Extremely blue bioluminewcence in the polychaete, *Polycirrus perplexus* (Terebellidae). *Bull.Mar.Sci.* 44: 1236-1239.

Hughes, L.E., and Lowry, J.K. (2015). A review of the world Cyphocarididae with description of three new species (Crustacea, Amphipoda, Lysianassoidea). *Zootaxa* 4058 (1): 001–040.

Johnson, F.H., Stachel, H.D., Shimomura, O., and Haneda, Y. (2015). Partial Purification of the Luminescence System of a Deep-sea Shrimp, *Hoplophorus gracilorostris.* In: *Bioluminescence in Progress,* pp.523-532.

Johnson, S., Frank, T.M., Haddock, S.H.D., Widder, E.A., Messing, C.G. (2012). Light and vision in the deep-sea benthos: I. Bioluminescence at 500–1000 m depth in the Bahamian Islands. *J.Exptl.Biol.* 215:3335-3343.

Johnsen, https://oceanexplorer.noaa.gov/explorations/15biolum/background/biolum/biolum.html

Jones, A., and Mallefet, J. (2010). Aposematic use of bioluminescence in *Ophiopsila aranea* (Ophiuroidea, Echinodermata). *Luminescence* 25: 81–216.

Jones, A., and Mallefet, J. (2012). Study of the luminescence in the black brittle-star *Ophiocomina nigra*: toward a new pattern of light emission in ophiuroids. *Zoosymposia* 7: 139-145.

Kanda, S. (1939). The Luminescence of a Nemertean, *Emplectonema kandai,* Kato. *Biol. Bull.* 77: 166-173.

Kates, H.S.C. (2011). *The Evolution of Luminescent Courtship Signals in Caribbean Ostracods.* Research Honors Thesis, Cornell University, https://hdl.handle.net/1813/23177

Kin, I., and Oba, Y. (2020). Bioluminescent properties of *Mesochaetopterus japonicus* (Polychaeta: Chaetopteridae) with comparison to *Chaetopterus. Plankton Benthos Res.*15: 228–231.

Kin, I., Jimi, N., and Oba, Y. (2019). Bioluminescence properties of *Thelepus japonicus* (Annelida:Terebelliformia). *Luminescence* 34:602–606.

Kin, I., Jimi, N., Mizuno, G., Koike, H., and Oba, Y. (2021). Bioluminescence of the polychaete *Tharyx* sp. (Annelida: Cirratulidae) in deep-seawater from Toyama Bay, Japan., *Plankton Benthos Res.* 16: 145–148.

Klein, J.C.V. (Ed.) (2000). *The Biodiversity Crisis and Crustacea*- Proceedings of the Fourth International Crustacean Congress. CRC Press, Science, 304 p.

Kotlobaya, A. (+18) (2019). Bioluminescence chemistry of fireworm *Odontosyllis. Proc. Natl. Acad. Sci. U.S.A.* 116: 201902095.

Latz, M. I., Frank, T. M. and Case, J. F. (1988). Spectral composition of bioluminescence of epipelagic organisms from the Sargasso Sea. *Mar. Biol.* 98: 441- 446.

León-González + 7 (2021). Anélidos Marinos de México y América Tropical. http://eprints.uanl.mx/22161/19/22161-1.pdf

Mallefet, J., Duchatelet, L., and Coubris, C. (2020). Bioluminescence induction in the ophiuroid *Amphiura filiformis* (Echinodermata) *J. Exp. Biol.* 223 : jeb218719.

Mallefet, J., and Isson, M.D.U. (1995). Preliminary results of luminescence control in isolated arms of *Ophiopsila aranea* (Echinodermata). *Belg. J. Zool.* 125 :167-173.

Markova, S.V., and Vysotski, E.S. (2015). Coelenterazine-Dependent Luciferases. *Biochemistry* 80: 714-732.

Martini, S., and Haddock, S.H.D. (2017). Quantification of bioluminescence from the surface to the deep sea demonstrates its predominance as an ecological trait. *Sci.Rep.* 7 (45750). DOI:10.1038/srep45750

Martini, S., Schultz, D.T., Lundsten, L., and Haddock, S.H.D. (2020). Bioluminescence in an Undescribed Species of Carnivorous Sponge (Cladorhizidae) From the Deep Sea. *Front. Mar. Sci.* https://doi.org/10.3389/fmars.2020.576476

Martini, S., Kuhnz, L., Mallefet, J., and Steven H. D. Haddock, S.H.D. (2019). Distribution and quantification of bioluminescence as an ecological trait in the deep sea benthos. *Sci Rep.* 9, 14654, https://doi.org/10.1038/s41598-019-50961-z

Mirza, J., and Oba, Y. (2021). Semi-Intrinsic Luminescence in Marine Organisms. DOI: 10.5772/intechopen.99369, 2021

Moraes, G.V., Hannon, M.C., Sores, D.M.m., Stevani, C.V., Schultze, A., Oliviera, A.G. (2021). Bioluminescence in Polynoid Scale Worms (Annelida: Polynoidae). *Front. Mar. Sci.* https://doi.org/10.3389/fmars.2021.643197

Morin, J.G., and Cohen, A.C. (1988). Two new luminescent ostracods of the genus *Vargula* (Myodocopida: Cyprinidae) from the San Blas region of Panama. *J. Crust.Biol.* 8: 620-638.

Muzik, K. (1978). A bioluminescent gorgonian, *Lepidisis olapa,* new species Coelenterata Octocorallia, from Hawaii. *Bull.Mar.Sci.* 284: 735-741.

Nicol, J.A.C. (1958a). Observations on the luminescence of Pennatula phosphorea, with a note on the luminescence of *Virgularia mirabilis. J.mar. bio.Ass.UK,* 37:551-563.

Nicol, J.A.C. (1953). Luminescence in polynoid worms. *J.mar.bio.Ass.UK* 32:65-84.

Nicol, J.A.C. (1958). Observations on luminescence in pelagic animals. *J.mar. bio.Ass.UK,* 37: 705-752.

Nishigushi, M.K., lopez, J.E., and Boletzky, S.V. (2004). Enlightenment of old ideas from new investigations: more questions regarding the evolution of bacteriogenic light organs in squids. *Evol. Dev.* 6: 41-49.

Nowel, M.S., Shelton, P.m., and Herring, P.J. (1998). Cuticular Photophores of Two Decapod Crustaceans, *Oplophorus spinosus* and *Systellaspis debilis. Biol Bull.* 195:290-307.

O'Reilly, S.C. (2016). *Identification of Deep-Sea Asteroidea and Ophiuroidea (Phylum Echinodermata) from two of Ireland's Submarine Canyon Systems, by use of Morphological and Molecular Techniques.* B.Sc. (Hons.) Marine Science Thesis, Murdoch University.

Oba, Y., Stevani, C.V., Oliviera, A.G., Tsarkova, A.S., Chepurnykh, T.V., Yampolsky, I.V. (2017). Selected Least Studied but not Forgotten Bioluminescent Systems. J. *Photochem.Photobiol.* 93: 405-415.

Oba, Y., Kato, S., Ojika, M. and Inouye, S. (2007). Biosynthesis of *Cypridina* Luciferin in *Cypridina noctiluca. Heterocycles* 72:10.3987/COM-06-S(K)27.

Odontosyllis polycera. Wikipedia

Ohmiya, Y., Applications of Bioluminescence. Cell Based Assays and Imaging. http://photobiology.info/Ohmiya.html

Okanishi, M., Oba, Y., and Fujita, Y. (2019). Brittle stars from a submarine cave of Christmas Island, northwestern Australia, with description of a new bioluminescent species *Ophiopsila xmasilluminans* (Echinodermata: Ophiuroidea) and notes on its behaviour. *Raffles Bull. Zool.* 67: 421–439.

Oliviera, A.G., Amaral, D.T., Hannon, M.C. Schulze, A. (2021). First Record of Bioluminescence in a Sipunculan Worm. *Front. Mar. Sci.* 8:762706. doi: 10.3389/fmars.2021.762706

Osamu , S., Shoji , I., Toshio , G. (1975). The light-emitter in bioluminescence of the sea cactus *Cavernularia obesa. Chem. Lett.* 4: 247-248.

Pablo, M.S. (2020). https://dial.uclouvain.be/memoire/ucl/en/object/thesis%3A23003

Plyuscheva, M., and Martin, D. (2009). On the morphology of elytra as luminescent organs in scale-worms (Polychaeta, Polynoidae) *Zoosymposia* 2: 379–389.

Poupin J., Cussatlegras A.S., and Geistdoerfer P. (1999) Plancton marine bioluminescent : Inventaire documenté des espèces et bilan des formes les plus communes de la mer d'Iroise. 1-83

Pringgenies., D., and Jorgensen, J.M. (1994). Morphology of the Luminous Organ of the Squid *Loligo duvauceli* d'Orbigny, 1839. *Acta Zool.* 75:305 – 309.

Raddatz. J., Hansteen, D.T., Correa, M.L., and Ru¨ggeberg, A. (2011). Bioluminescence in deep-sea isidid gorgonians from the Cape Verde archipelago. *Coral Reefs* 30:579.

Reda, N.J., Morin, J.G., Torres, E., Cohen, A.C., Schawaroch, V., and Gerrish, G.A. (2019). *Maristella,* a new bioluminescent ostracod genus in the Myodocopida (Cypridinidae). *Zool. J. Linn. Soc.* 187: 1078-1118.

Reid, A.L. (2021). Two new species of *Iridoteuthis* (Cephalopoda: Sepiolidae: Heteroteuthinae) from the southwest Pacific, with a redescription of *Stoloteuthis maoria* (Dell, 1959), *Zootaxa* 5005: 503-537.

Robison, B.H. (1992). Bioluminescence in the benthopelagic holothurian *Enypniastes eximia. J.mar.bio.Ass.UK,* 72: 463 – 472.

Ropert, C.F.W. (1963). Observations on bioluminescence in *Ommastrephes pteropus* (Steenstrup, 1855), with notes on its occurrence in the Family Ommastrephidae (Mollusca: Cephalopoda). *Bull.Mar.Sci.* :13, 343-353.

Satterlie, R.A., and Case, J.F. (1979). Development of bioluminescence and other effector responses in the pennatulid coelenterate *Renilla köllikeri. Bio. Bull.* 157: 506-523.

Sanna, T., Dallolio, L., Raggi, A., Mazzetti, M., Lorusso, G., Zanni, A., Farruggia, P., and Leoni, F. (2018). ATP bioluminescence assay for evaluating cleaning practices in operating theatres: applicability and limitations. *BMC Infect. Dis.* 18: 583.

Schweikert, L.E., Davis, A.L., Johnson, S., and Bracken]Grissom, H.D. (2020). Visual perception of light organ patterns in deep]sea shrimps and implications for conspecific recognition. *Ecol Evol.*10: 9503–9513.

Shimomura, O., and Yampolsky, I.V. (2019). *Bioluminescence: Chemical Principles and Methods* (3rd Edition). World Scientific, - Science - 556 p.

Shimomura, O. (2008). Bioluminescence of the brittle star *Ophiopsila californica. Photochem. Photobiol.* 44:671 – 674.

Shimomura, O., and Johnson, F.H. (1975). Chemical nature of bioluminescence systems in coelenterates. *Proc.Nat.Acad.Sci.* USA, 72: 1546-1549.

Soaes, M. (2020). On the bioluminescence of Echinodermata in the south-eastern Australian deep-sea: New insights into sacculi and light emission in Comatulida (Crinoidea).

Faculté des sciences, Master en biologie des organismes et écologie.

Stokes, https://www.csmonitor.com/Science/2012/0906/On-ocean-floor-a-shrimp-that-vomits-light

Syed, A.J., and Anderson, J.C. (2021). Applications of bioluminescence in biotechnology and beyond. *Chem. Soc. Rev.* 50:5668-5705.

Taboada, S., Silva, A.S., Diez-Vives, C., Neal, L., Cristobo, J., Rios, P., Hestetun, J.T., Clark, B., Ross, M.E., Junoy, J., Navarro, J., and Riesgo, A. (2020). Sleeping with the enemy: unravelling the symbiotic relationships between the scale worm *Neopolynoe chondrocladiae* (Annelida: Polynoidae) and its carnivorous sponge hosts. *Zool. J. Linn. Soc.* 193:295-318.

Tilic, E., and Rouse, G.W. (2020). Hidden in plain sight, *Chaetopterus dewysee* sp. nov. (Chaetopteridae, Annelida) – A new species from Southern California. *Eur. J. Taxon.* 643: 1–16.

Tinikul, ., Chunthaboon, P., Phonbupph, J., Paladkong, T. (2020). Bacterial luciferase: Molecular mechanisms and applications. *Enzyme* 47: 427-455.

Torres, E., and Cohen, A.C. (2005). *Vargula morini*, a New Species of Bioluminescent Ostracode (Myodocopida: Cypridinidae) from Belize and an Associated Copepod (Copepoda: Siphonostomatoida: Nicothoidae). *J.Crust.Biol.* 25: 11-24.

Torres. E. (2009). *Vargula Annecohenae*, a New Species of Bioluminescent Ostracode (Myodocopida: Cypridinidae) from Belize. *J.Crust.Biol.* 27: 649-659.

Tsuji, F.I. (1983). Molluscan bioluminescence. In: P.W. Hochachka (Ed.), *Environmental Biochemistry and Physiology, The Mollusca*, vol. 2, Academic Press, New York, pp. 257 – 279

Tsuji, A., Stanley, P.E., Kricka, L.J., Maeda, M., Matsumoto, M. (2005). Bioluminescence and Chemiluminescence: Progress and Perspectives. *Proc. 13th Int. Symp.* 596 p.

Valle's Y., and Gosliner, T.M. (2006). Shedding Light onto the Genera (Mollusca: Nudibranchia) *Kaloplocamus* and *Plocamopherus* with Description of New Species Belonging to These Unique Bioluminescent Dorids, *Veliger* 48:178–205.

Vargulin. https://en.wikipedia.org/wiki/Vargulin

Verdes, A., and Holford, M. (2018). Beach to Bench to Bedside: Marine Invertebrate Biochemical Adaptations and Their Applications in Biotechnology and Biomedicine. *Results Probl. Cell Differ.* 65: 359-376.

Verdes, A., Alvarez-Campos, P., Nygren, A., Martin, G.S., Rouse, G., Deheyn, D.D., Gruber, D.F., and Holford, M. (2018). Molecular phylogeny of *Odontosyllis* (Annelida, Syllidae): A recent and rapid radiation of marine bioluminescent worms. *bioRxiv* preprint doi: https://doi.org/10.1101/241570;

Verdes, A., Gruber, D.F. (2017). Glowing Worms: Biological, Chemical, and Functional Diversity of Bioluminescent Annelids. *Integr. Comp. Biol.* 57: 18–32.

Verdes *et al.*, https://www.readcube.com/articles/10.1101/241570

Vestheim, H., and Kaartvedt, S. (2009). Vertical migration, feeding and colouration in the mesopelagic shrimp *Sergestes arcticus. J. Plankton Res.* 31: 1427-1435.

Wang, X., Fan, X., Schröder, H.C. *et al.* (2012). Flashing light in sponges through their siliceous fiber network: A new strategy of "neuronal transmission" in animals. *Chin. Sci. Bull.* 57: 3300–3311.

Wei, S.L., and Young, R.E. (1989). Development of symbiotic bacterial bioluminescence in a nearshore cephalopod, *Euprymna scolopes. Mar.Biol.* 103: 541-546.

Widder, E.A., Latz, M.I., and Case, J.F. (1983). Marine bioluminescence spectra measured with an optical multichannel detection system. *Biol. Bull.* 165: 791-810.

Wilson, M.F. (2016). The science behind bioluminescence in the sea. https://www.juneauempire.com/life/the-science-behind-bioluminescence-in-the-sea/

Zaragoza, N., Quetglas, A., Moreno, A. (2015.Identification guide for cephalopod paralarvae from the Mediterranean Sea. *ICES Cooperative Research Report*, Vol. 324, 97 p.

Zorner, S., and Fischer, A. (2006). The spatial pattern of bioluminescent flashes in the polychaete *Eusyllis blomstrandi* (Annelida). *Helgol Mar Res* 61:55-66.

Index

B

C

D

E

J

K

L

M

N

O

P

R

S

T

U

V

W

Z

www.ingramcontent.com/pod-product-compliance
Ingram Content Group UK Ltd.
Pitfield, Milton Keynes, MK11 3LW, UK
UKHW021010290726
14059UKWH00001BA/60